CLONING
Nuclear Transplantation in Amphibia

Frontispiece. An early, if not the earliest, illustration of the frog species first used in successful nuclear transplantation experiments. The illustration is reprinted from Mark Catesby, *The Natural History of Carolina, Florida, and the Bahama Islands: Containing Figures of Birds, Beasts, Fishes Serpents, Insects and Plants, not Hitherto Described, or very Incorrectly Figured by Authors. Together with their descriptions in English and French. To which are added, Observations on the Air, Soil, and Waters: With remarks upon Agriculture, Grain, Pulse, Roots, etc. To the whole is prefixed a new and correct map of the countries treated of,* Volume II, large folio (London, 1731), plate 70.

Catesby's description of the frog, after almost two and a half centuries, remains essentially correct, although I do take exception to the claim that the frog "will leap at once five or six yards." Mark Catesby's characterization of the frog's nimbleness is not inconsistent with Mark Twain's assertions concerning a famous jumping anuran from Calaveras County. My field observations are not in harmony with either of the Marks, Catesby or Twain.

CLONING
NUCLEAR TRANSPLANTATION
IN AMPHIBIA

A Critique of Results
Obtained with the Technique
to Which Is Added
a Discourse on the Methods
of the Craft

Robert Gilmore McKinnell

Professor of Genetics and Cell Biology
College of Biological Sciences,
University of Minnesota

UNIVERSITY OF MINNESOTA PRESS ☐ MINNEAPOLIS

QH
442.2
.M33
1978

Appendix V, Sources of Amphibians for Research II, copyright © 1978
by George W. Nace and Jeffrey K. Rosen.

The cover illustration of a sexually mature nuclear transplant frog
and progeny is from R. G. McKinnell, Intraspecific nuclear transplantation
in frogs, *J. Hered.*, 1962, 53, 199-207, Figure 3.

This book
is for the mother
of
Nancy Elizabeth, Robert Gilmore, and Susan Kerr

Preface

A note concerning the title of this book seems to me to be appropriate. Biologists generally use the word "cloning" to refer to the production of multiple genetically identical individuals. Cloned amphibia, in the sense of isogenic groups of frogs, are produced by nuclear transplantation, which is the subject of this book. However, the word "cloning" is used here in a more general sense, that is, to refer to *one* or more individuals produced by the method. A paragraph from Kass (1972) is reproduced here because it is an enlightened discourse on the etymology and meaning of the word.

Some people prefer to reserve the use of the term "cloning" for (such) mass-scale replication and prefer instead to use "nuclear transfer" to describe the single instances of asexual reproduction, hoping thereby not to bring the opprobrium of the mass use upon individual cases. I find this suggestion conceptually fuzzy and etymologically problematic. "Clone" comes from the Greek word *klōn*, meaning "twig" and is related to the word *klan*, meaning "to break." It is defined (*Webster's Third International Dictionary*) as "the aggregate of asexually produced progeny of an individual, whether natural (as the products of repeated fission of a protozoan) or otherwise (as in the propagation of a particular plant by budding or by cuttings through many generations)." While the term *clone* does imply an aggregate, it is an aggregate which is formed not horizontally (during one generation) but vertically over time (generation after generation). Thus even the first asexually-produced offspring and his progenitor together form a clone, albeit a small one for the time being.

"Nuclear transfer" is but the name of one of several possible techniques that could give rise to a clone and thus does not serve as a generic term for the genetically and humanly significant features of asexual reproduction. Moreover, the desire to avoid for the small-scale use the offensive connotation of mass production begs the question of whether the opprobrium is not equally fitting and leads towards the development of euphemism. I shall use "cloning" as synonymous with "asexual reproduction, artificially induced."

My reasons for writing this book relate to at least three situations in my life. The first has to do with classroom teaching. On several occasions in the past few years I have offered a seminar in amphibian nuclear transplantation. I have handed out class notes to supplement my lectures. I thought it would be useful to have a book for the students instead of the notes, which did not contain illustrations and which were easily lost. Another reason for this writing concerns teaching the cloning procedure to students and other colleagues. I hope the appendixes are helpful in showing them how to transplant nuclei. Readers of the first seven chapters will better understand the limitations and opportunities of the procedure by referring to the several appendixes. Finally, comments are made from time to time which reveal how little even professional biologists know about amphibian nuclear transfer results and procedures. When such comments are made to me in the future, I shall turn to an appropriate section of the book to enlighten my friends.

Dr. Thomas J. King, Director, Division of Cancer Research Resources and Centers, National Cancer Institute, introduced me to the nuclear transplantation procedure in his laboratory at the Institute for Cancer Research, Philadelphia. Dr. Marie A. DiBerardino, Professor of Anatomy, Medical College of Pennsylvania, Philadelphia, a long-time colleague and friend, read the entire manuscript and made many valuable comments. I am in her debt. Dr. D. J. Kilian, Director, Biomedical Research and Environmental Health, Dow Chemical U.S.A., Texas Division, gave encouragement to the author at a particularly important time. Lyle M. Steven, Jr., read the manuscript several times and was exceptionally helpful. John W. Schaad IV sought and found references. Janet Sauer and Thomas Fontaine, Jr., assisted in reading proof. However, errors undoubtedly were introduced as the manuscript evolved and the errors are of course mine. Beverly Kerr McKinnell helped with the bibliography and in other ways and I thank her.

I have been the recipient of research funds from the National Cancer Institute, the National Science Foundation, the American Cancer Society, Inc., the Damon Runyon Memorial Fund for Cancer Research, Inc., Dow Chemical U.S.A., and the Graduate School of the University of Minnesota. The grants, for which I am greatly appreciative, supported my nuclear transplantation studies and indirectly made this volume possible.

I hope that W. P. M. sees this effort. I wish that M. C. G. could have. I thank the former for the capacity to endure and I thank the latter for the desire to achieve—and I thank both for a warm and happy home so many years ago.

R. G. M.

St. Paul, Minnesota
March 1978

Table of Contents

xi

Appendixes

CLONING
Nuclear Transplantation in Amphibia

"Il y a dans ces opérations une mortalité notable."

Gallien, Picheral, and Lacroix, 1963a

CHAPTER 1

Some Early Experiments That Suggested the Feasibility of Amphibian Cloning

Nuclear transplantation is a technique designed to provide information answering various biological questions. The procedure was initially addressed to the question of totipotency of somatic nuclei. That question is related to the germ plasm theory of Weismann (discussed below) and to the results of the separation of blastomeres in marine and amphibian embryos. It is instructive, therefore, to review the experiments that preceded the first cloning studies of Briggs and King (1952).

Subsequent chapters will be concerned with the exploitation of the cloning procedure in amphibians. However, nuclear transfer in protista will be considered in the present chapter because historically the tools of cloning (Appendix II) evolved, at least in part, from the tools of the individuals who studied amoebae nuclear grafting.

The history of the formation of our concepts of cell differentiation can be discouraging because the very questions posed by pioneers remain unanswered (but relevant) to this day (Oppenheimer, 1965). Experiments relating to the role of the nucleus in development, preceding the nuclear transfer experiments in frogs, are reviewed in detail by Wilson (1928), Morgan (1927), and Spemann (1938).

3

Spemann's Views on a Postulated Experiment
That "Appears, at First Sight, to Be
Somewhat Fantastical"

From the assumption of Weismann concerning the differential division of the nucleus, there would follow *immediately a restriction of the potency of the genome*; for if the germ plasm were separated, during development into its constituent parts and distributed over the single cells of the body, it is evident that each of these cells would contain only its allotted portion and would lack all the remainder. But, by refuting of this fundamental hypothesis one cannot conclude that each cell of the body will contain the whole undiminished idioplasm; for genes may be lost or *become ineffective in other ways besides that of elimination out of the cell.* Decisive information about this question may perhaps be afforded by an experiment which appears, at first sight, to be *somewhat fantastical.* It has been shown, as pointed out before, in the egg of the sea urchin (Loeb, 1894) and the newt (Spemann, 1914) that a piece of the egg protoplasm which contains no nucleus may be induced to develop, may be "fertilized," as it were by a descendant of the fertilized egg nucleus. In this experiment the nucleus to be tested is transmitted through a bridge of protoplasm from the one half of the egg in which it has originated into the other half which had hitherto been devoid of a nucleus. *Probably the effect could be attained if one could isolate the nuclei of the morula and introduce one of them into an egg or an egg fragment without an egg nucleus.* The first half of this experiment, to provide an isolated nucleus, might be attained by grinding the cells between two slides, whereas for the second, the introduction of an isolated nucleus into the protoplasm of an egg devoid of a nucleus, I see no way for the moment. If it were found, the experiment would have to be extended so that older nuclei of various cells could be used. *This experiment might possibly show that even nuclei of differentiated cells can initiate normal development in the egg protoplasm* (Spemann, 1938, pp. 210-211) (italics added).

Hans Spemann (Figure A-1), who wrote the above paragraph, was the first embryologist to receive the Nobel Prize for research in developmental biology. The fact that few Nobel Prizes have been conferred upon embryologists is not a result of failure to ask appropriate questions but reflects on the difficulty of obtaining answers. Spemann thought that a nucleus of a morula might provide the genetic program for an entire embryo. He hoped for a method to introduce a morula nucleus into a nucleusless egg—the wish was fulfilled within a decade and a half. Spemann wondered if the nucleus of a differentiated cell could start embryonic development in egg cytoplasm. If it failed to do so, the failure would not necessarily be due to an incomplete genome (i.e., failure of a differentiated cell nucleus to promote normal development when transplanted to an egg devoid of a nucleus would not necessarily disprove that such nuclei have informational equivalence to the zygote nucleus), but the failure presumably could be due to other factors such as differential gene activity that is highly stabilized in differentiated cells. Note that Spemann considered the possibility that genes may "become ineffective in other ways besides that of elimination out of the cell." Speman's phraseology suggests that he believed some genetic loci may be active while others are present but repressed, an idea in harmony with modern concepts regarding gene regulation (Stein, Spelsberg, and Kleinsmith, 1974; Ford, 1976).

If it could be clearly established that informational equivalence obtains in adult nuclei, then the way would be clear to ask, by means of other kinds of experiments, what controls the differentiation of cells with genetically identical nuclei. Questions relating to nuclear control of cytoplasmic activity have been important to embryologists throughout the twentieth century. Cytoplasmic control of nuclear activity, the other side of the coin, has been studied equally as long.

Resolution of questions regarding nucleocytoplasmic interactions was sought primarily through manipulative procedures. Although the second half of this century has witnessed truly marvelous and exciting revelations of biochemistry and molecular biology, it should be noted at this point that there is yet no way for a biochemist to examine a blastomere and predict whether or not, if isolated, it will develop into a whole or partial embryo. The manipulative embryologists, who were called microsurgeons in Spemann's day, still provide the only means to an answer to that and certain other similar questions.

Weismann's Germ Plasm

August Weismann's germ plasm theory (1892) was considered

by Spemann (1938) to be one of the major studies inaugurating the experimental analysis of development. Weismann observed that a zygote formed from the union of sex cells has a developmental potentiality far greater than that normally expressed by any adult somatic cell. Primordial sex cells are segregated early in many invertebrates and are conspicuous in early chordate embryos (Blackler, 1970). What accounts for the omnipotency of gametes descended from primordial germ cells and the restricted capabilities of somatic cells?

Weismann erected a landmark theory that, although it has been entirely abandoned, stimulated much thought in the last decades of the nineteenth century and the early years of this century. Weismann believed that differentiation was controlled by determinants. The determinants were unique for each cell type and they controlled the structure and function of each cell type. The determinants were derived by partitioning the genome (the ids) through mitosis. The genome was subdivided or partitioned only in the somatic nuclei. The cells of the germ plasm retained the intact genome (the idioplasm).

Chromosomes were thought to be aggregates of ids. "In the first cell division every id divides into two halves, each of which contains only half the entire number of determinants, and this process of disintegration is repeated at every subsequent cell division, so that the ids of the following ontogenetic stages gradually become poorer as regards the diversity of their determinants, until they finally contain only a single kind" (Weismann, 1892).

The germ line and its special cytoplasm are still being studied many years after Weismann published his now discarded theory. The "germinal cytoplasm" has been scrutinized by light microscopy (DiBerardino, 1961) and by electron microscopy (Mahowald and Hennen, 1971). Experimental studies include Smith (1966), Mims and McKinnell (1971), Reynaud (1973), and Illmensee, Mahowald, and Loomis (1976). Briefly, it can be shown that the early segregating primordial germ cells are de facto germ cells not because of nuclear peculiarities but because of special cytoplasm (at least in *Drosophila* and in certain frogs) that is rich in mitochondria and stains as does RNA. Thus, Weismann, who thought that germ cells were different from somatic cells because of an undiminished genome, was wrong. The uniqueness of germ cells must be attributed to *cytoplasmic* determinants.

An obvious experimentally testable inference followed from Weismann's germ plasm theory. Qualitative differences must be found among the progeny of a zygote nucleus. Potency of isolated blastomeres should be less than the potency of the intact embryo. Wilhelm Roux's experiment with frog eggs seemed to support Weismann's idea.

Half Embryos after Injury to a Blastomere

Roux (1888) injured one blastomere of a two-cell frog embryo (*Rana esculenta*) by plunging the tip of a hot needle into it. The wounded cell remained in place next to the normal blastomere. A half embryo developed (Figure 1-1). Roux judged from this and other experiments that "cleavage divides qualitatively that part of the embryonic, especially the nuclear material that is responsible for the direct development of the individual by the arrangement of the various separated materials which takes place at that time, and it determines simultaneously the position of the later differentiated organs of the embryo." It would seem reasonable that one blastomere of a two-cell stage embryo developed into a half embryo because it contained only half of the determinants for a

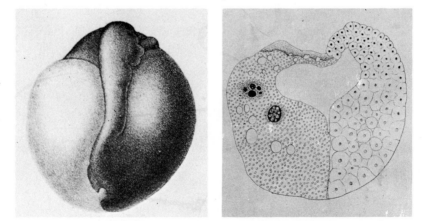

Figure 1-1. (*left*) A frog half embryo that is the result of an injury to one of the first two blastomeres. (*right*) Histological section of a half blastula produced in the same manner. (From Roux, 1888.)

whole embryo. Thus, Weisman regarded Roux's experiment as support for his theory of the germ plasm.

Roux's results with frog embryos were not dissimilar to those of Chabry (1887) and Conklin (1905) with ascidian embryos. Chabry damaged blastomeres with a glass needle and Conklin effected the same end by squirting dividing ova in and out of a dropping pipette. Both recorded imperfect development associated with loss of one or more blastomeres. While a mosaic development was described, it should be noted here that defective embryos frequently are competent to restore or regenerate ("post-generation") many missing parts.

Chabry's early experiments were an incentive perhaps for those of Driesch (1892a) who separated two-cell sea urchin (*Echinus microtuberculatus*) embryos by shaking them violently in sea water. Driesch's separated blastomeres developed into normal appearing but diminutive larvae (Figure 1-2). Larvae were also obtained from blastomeres of the four-cell stage embryo. One blastomere from an eight-cell sea urchin is occasionally competent to form a miniature pluteus larva (Hörstadius and Wolsky, 1936; see also Hörstadius, 1973). Driesch wondered if the results he reported might not also obtain in the frog if the blastomeres were physically separated. He lamented: "I have tried in vain to isolate amphibian blastomeres; let those who are more skillful than I try their luck."

O. Hertwig (1893) tried his luck with a hair-loop constriction of a two-cell embryo. He failed. But, more skillful indeed (or perhaps, more persistent) were Endres (1895) and Spemann (1901, 1938) who repeated Driesch's experiments with urodele embryos and Brachet (1904) and Schmidt (1933) who worked with anuran ova. If the gray crescent material is evenly distributed between the two separated daughter blastomeres, two perfect but small embryos result—a finding not at all harmonious with Weismann's hypothesis.

Equivalence of blastomeres in forms other than sea urchins and amphibians was demonstrated in those early years of experimental embryology. A normal hydroid develops from an isolated blastomere of a two-cell or a four-cell stage *Clythia flavidula* (Zoja, 1895). A whole amphioxus embryo develops from a singe cell of a

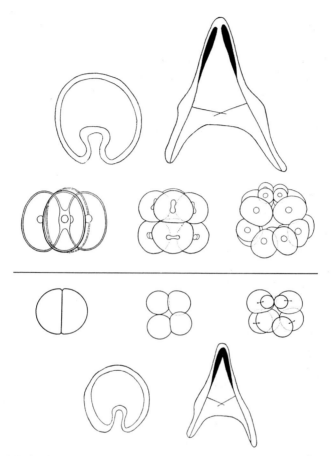

Figure 1-2. (*top*) Normal cleavage, morula, gastrula, and pluteus larva of the sea urchin. (*bottom*) Similar but miniature developmental stages of the sea urchin following isolation of a blastomere at the two-cell stage. (From Driesch, 1892a.)

two-blastomere embryo (Wilson, 1893). Similar results were obtained with the nemertean *Cerebratulus* (Wilson, 1903) and teleost fishes (Morgan, 1895). Spirally cleaving eggs (to be discussed below) did not yield comparable results during those halcyon days of manipulative embryology.

Before proceeding further, it would be instructive to emphasize that equivalence in the earliest cleavages in the forms described is invariably followed by loss of omnipotentiality after only one or two mitotic divisions. Thus, while an occasional 1/8th sea urchin may become a small pluteus (Hörstadius and Wolsky, 1936), most individuals developing as isolated cells from the eight-blastomere stage develop no further than the swimming blastula. The 1/8th blastomere of the hydroid *Clythia* forms a hollow blastula, fills with endoderm, and develops no further. Few amphioxus embryos are formed from separated blastomeres of the four-cell embryo. Imperfect and arrested development ensue from 1/4th embryos of *Cerebratulus*. *Fundulus* 1/4th embryos are probably all defective. It is generally thought that incomplete development associated with blastomeres from the second, third, and fourth cleavages is probably due to quantitatively inadequate cytoplasmic materials.

The early embryologists noted that when an entire embryo was formed from a blastomere, it was invariably smaller than normal embryos formed from the entire ovum. The observers recorded that although the experimental embryo was small, its constituent cells were normal in size. Thus, a whole embryo was structured with few cells which were of relatively enormous size. What imposes the few cells to form a whole embryo and what enjoins mitotic activity and rates of cell synthesis to organize the miniature "perfect" embryo? These questions are unanswered.

Other experiments suggested the interchangeability of nuclei and thus the equivalence of genomes. Schultze (1895), without separating cells, demonstrated the equipotentiality of blastomeres by inverting the two-cell embryo. Monsters of varying degrees of doubling developed from the topsy-turvy experiment. Cleavage under pressure between plates of glass causes a displacement of nuclei such that they come to reside in cytoplasm that they would not otherwise inhabit. Normal development follows when the forces of compression are released (Driesch, 1892b). The cited experiments prove that during cleavage the nuclei divide in such a way that their daughter-nuclei are quantitatively and qualitatively equal. Similar conclusions are justified from the observations of fused embryos. One large embryo results (or alternatively, two joined embryos may develop depending on the position of the cleavage furrow with respect to the gray crescent) (Mangold and Seidel, 1927).

Nuclear equivalence in insects is attested to by the results of ultraviolet irradiation of one nucleus at the two-nuclei stage of the dragonfly, *Platycnemis*. Normality follows despite loss of 50% of nuclear material at an early age (Seidel, 1932). As could be predicted, the early studies in insects were followed by nuclear transplantation studies in the fruit fly, *Drosophila*, which confirm the genomic equivalence of early division nuclei (Illmensee, 1973).

The ancient and venerable experiments cited here argue effectively that there is an equivalence in developmental potentialities among nuclei of early stages in *selected* embryos, notably chordates, hydroids, nemertians, certain insects, and sea urchins. Ova of these species accordingly have been referred to as "indeterminate" or "regulative."

Technical procedures are now such that one can show that isolated mammalian blastomeres can give rise to a whole embryo. The procedure is to remove a developing egg from its mother, kill one or more of the early blastomeres with puncture by a microneedle, then implant the operated egg into the oviducts of a foster mother. An entire young animal is born later, thus showing the developmental totipotency of the single blastomere (Figure 1-3). This striking repeat of the late nineteenth- and early twentieth-century classic experiments made with oviparous animals has been described for the mouse (Tarkowski, 1971) and the rabbit (Moore, Adams, and Rowson, 1968). Thus, we now have laboratory experimental evidence that supports our previous notions derived from observations of natural phenomena such as twinning, that the mammalian egg is in fact "regulative" and undergoes "indeterminate" cleavage.

Quite dissimilar results were reported with eggs and embryos of ctenophores, rotifers, nematodes, ascidians, and other groups. Prototypic experiments with ova of these groups revealed little interaction between blastomeres, a reduced ability to regenerate missing parts of the early larva, and a high degree of predictability of furrow formation and position. Such development is referred to as "determinant"or "mosaic."

In contrast to indeterminate cleavage exemplified by the frog, consider if you will the development of isolated blastomeres of the ascidian egg referred to earlier. The scholarly contributions of Chabry (1887), and Conklin (1905) are summarized by Morgan

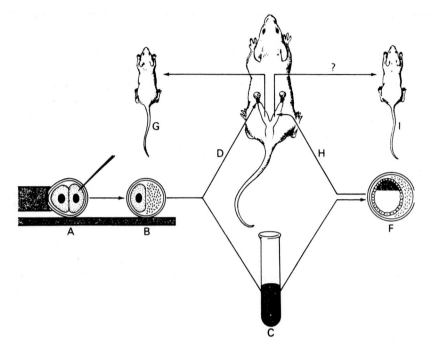

Figure 1-3. Scheme of an experiment that demonstrates totipotency of a mammalian blastomere at the two-cell stage. (A) One blastomere is injured with a needle. The surviving blastomere (B) may be cultured (C) to the blastocyst stage (F), or it may be placed in the reproductive tract of a foster mother (D) where it can develop to birth (G). The cultured blastocyst (F) is thought to be sufficiently normal that it too could be placed in a foster mother (H) where growth to term could occur (I). (Drawing by J. Modlinsky from Tarkowski, in *Methods in Mammalian Embryology*, edited by Joseph C. Daniel, Jr. W. H. Freeman and Company © 1971.

(1927): "In the plain meaning of the terms 'whole' and 'half,' there can be no question but that the 1/2 isolated blastomeres give rise to a half-embryo."

Driesch and Morgan (1895) recorded that the eight-combed sea jelly (ctenophora) developed as four-combed embryos when derived from a blastomere of the two-cell stage, and embryos have only two combs when they develop as 1/4th embryos. Mosaic development has been shown to occur in molluscs (Wilson, 1904) and in annelids (Penners, 1926). One should not be surprised, therefore, to learn that Spemann (1938, p. 210) stated, when contrasting mosaic and regulative eggs: "the processes of differentiation do not seem to me to be directly comparable in these two kinds of eggs."

Spemann was cheated by death from seeing what transpired during the latter part of the twentieth century. Had he been so long lived as to have persisted, perhaps he would have deleted the above phrase from his book. It seems that now, more than ever before, there is no reason to postulate fundamental differences between mosaic and regulative eggs.

To support this point of view, note that deletion of one blastomere at the two-cell stage of the spirally cleaving acoel, *Childia*, results in normal or near-normal larvae half the regular size (Boyer, 1971). Ascidian eggs develop as mosaics as far as their isolated blastomeres are concerned. But, unfertilized eggs of ascidians are "regulative." The virgin ova of *Ascidia malaca* may be sectioned into two fragments, fertilized, and *both* parts give rise to well-formed embryos (Reverberi and Ortolani, 1962). Isolation of first cleavage blastomeres reveals a strictly mosaic development of the gastropod *Bithynia tentaculata*. However, isolation of half embryos at the four-cell stage of this spirally cleaving embryo can be done in such a manner that both halves form all adult structures (Verdonk and Cather, 1973). Half embryos produced at the two- and four-cell stage in another spirally cleaving mollusc can regulate and develop into normal snails (Morrill, Blair, and Larsen, 1973). The above experiments demonstrate that at least certain eggs thought to be "mosaic" can in fact be shown to be "regulative." The point in discussing these instances here is that it seems to me there is little merit in postulating different control mechanisms among eukaryotic cells. All concepts regarding control of gene expression have the prerequisite of equivalence of genetic content of somatic nuclei. I believe that the blastomere separation studies available as of this writing support the view that there are probably no fundamental differences between determinant and indeterminate eggs and the molecular mechanisms underlying differentiation are likely to be the same among all nucleated (eukaryotic) organisms.

The discussion above concerns only the earliest embryonic events—what was known concerning later stages?

Delayed Nucleation Experiments of Loeb and Spemann

A primordial style of nuclear transplantation was performed by the one who referred to the procedure as "fantastical." The delayed nucleation experiments of Hans Spemann (summarized in Spemann, 1938) are indeed manipulatively sophisticated pioneer cloning experiments. Spemann's elegant studies were preceded by manipulatively different but essentially identical experiments with echinoderm ova.

Sea urchin zygotes insulted with hypotonic saline solution tend to swell. The vitelline membrane will respond to the affront by a break, which permits the formation of a cytoplasmic exudate or bleb. Under these circumstances, the herniated egg cytoplasm initially lacks a nucleus and therefore fails to develop for a time. However, after a few nuclear divisons of the nonherniated egg half, a daughter nucleus sometimes moves to the cytoplasmic exudate. When this happens, two whole embryos or a double monster develops, one from each portion of the herniated egg (Loeb, 1894).

The cytoplasm of a fertilized newt egg can be divided partially in a manner such that it develops similarly to the ruptured sea urchin eggs of Loeb. Spemann constricted newt zygotes with baby hair ligatures before the first division of the zygote nucleus. The stricture formed a dumbbell-shaped egg with the cytoplasmic bodies connected by a narrow isthmus. The zygote nucleus was allowed to divide and the part in which it divided became nucleated and cleaved while the non-nucleated portion remained an acellular mass not unlike an enucleate merogone. During the 8- or 16-cell stage of the dividing portion, one of the daughter nuclei (diploid but comprising only 1/8th or 1/16th of the nuclear progeny of the zygote) was allowed to traverse the isthmus and become resident in the previously non-nucleated cytoplasmic body. The latter portion, in a number of cases, developed into normal but younger and smaller larvae. It would seem therefore that a 1/8th or 1/16th descendant from a zygote nucleus contains sufficient genetic information to program the structuring and orderly sequence of events that result in larvae (Figure 1-4).

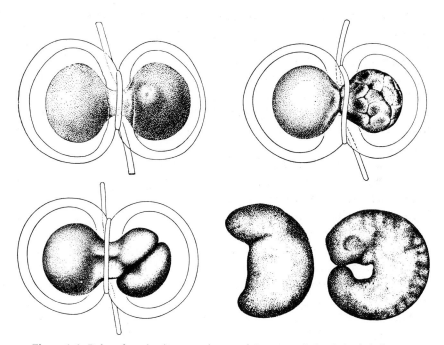

Figure 1-4. Delayed nucleation experiment of Spemann. (*A*) A baby hair ligature is applied to an undivided egg which results in a dumbbell-shaped egg. (*B*) First cleavage division is restricted to the egg portion that contains the zygote nucleus. At the morula stage of the constricted embryo (*C*), the ligature is released, a nucleus moves across to the previously non-nucleated half, and two embryos ensue (*D*). The embryo on the left is younger than the embryo on the right by an interval of time equivalent to three to five cleavage divisions. (From Spemann, 1938.)

Prognosticators

Spemann (see quotation at beginning of this chapter) was not alone in suggesting the value of a nuclear transfer procedure. King (1966) cites a number of workers who forecast the usefulness of such a technique, among them Rauber in 1886, Hämmerling in 1934, Lopashov in 1945, Lorch and Danielli in 1950, and Schultz in 1952. Success with nuclear transfer in amphibia was reported by Briggs and King (1952).

One of the more interesting seers of the pre-cloning era was

Jean Rostand (1943). He speculated in considerable detail on methodology of nuclear exchange (see Appendix II for present methods), wrote of fusing gamete nuclei with introduced nuclei resulting in polyploid embryos (Chapters 3 and 7), and considered the possibility of interspecific nucleocytoplasmic hybrids (Chapter 7). His concluding paragraph, a third of a century ago, says: "En effet, il serait bien intéressant de préciser le moment de l'ontogenèse—moment qui doit différer suivant les tissus—où les noyaux somatiques perdent la capacité de diriger le développment de l'oeuf. D'autre part, l'inoculation de noyaux embryonnaires dans un oeuf d'espèce différente, préalablement stérilisé par le radium, permettrait de réaliser la mérogonie hybride diploide (par exemple, inoculation d'un noyau embryonnaire de *Rana arvalis* dans un oeuf stérilisé de *Rana temporaria*)."

Nineteenth- and Early Twentieth-Century
Nuclear Transfer Experiments

Oppenheimer (1965) reminded some of us that nuclear transfer is not a young technique. While many speculated (see section above), a few turned to the laboratory bench. Thus, Rauber (1886) made intergeneric nuclear transfers. Oppenheimer stated that Roux (1892) was apprised of Rauber's study and asked: "Did Driesch know of these experiments?" Oppenheimer's answer is that he did not refer to them "nor have many others since then." Although Driesch (1894) failed to cite Rauber, he did know of the work at least at a later date. Driesch cited Rauber subsequently in a very brief manner, probably because he "failed to get any results at all" (Driesch, 1908, p. 235).

At almost the same time that Driesch wrote those words, it was claimed that nuclear exchange was fait accompli. Guyer (1907) pricked uterine eggs of frogs with a fine capillary pipette containing blood cells. Many blastulae and gastrulae developed. Two pricked eggs became swimming tadpoles. Guyer believed that his embryos resulted from proliferation of the introduced leucocytes and he was "inclined to believe that in some cases at least, the female pronucleus of the egg takes no part in the proliferation."

If Guyer were correct in his belief that tadpoles developed in response to the nuclear proliferation of introduced blood cells, then his experiment would be doubly significant. It would be of historical interest because it preceded by 45 years the first reports of Briggs and King (1952), but it is of even greater interest because of its relevance to the question of nuclear potentialities posed by Weismann and others. Guyer's study suggests that at least some cells derived from *adult* organisms retain substantial developmental potentialities and can initiate normal development in the egg cytoplasm. This was the ultimate type of experiment referred to by Spemann.

Because of these reasons, Guyer's experiment was repeated recently. The new experiment utilized polyploid (3n) blood cells in an effort to distinguish gynogenesis (1n or 2n development) from development initiated by introduced blood cells. The results cast doubt on Guyer's speculations because no triploid embryos (i.e., embryos of the ploidy of the introduced blood cells) resulted, although swimming haploid and diploid embryos were obtained. Thus, Guyer was observing parthenogenesis and not the effects of nuclear transfer. The recent experiments with polyploid blood cells are described in Chapter 4.

Amoeba Nuclear Transfer. How can a discussion of nuclear transfer occur in a chapter about the feasibility of nuclear transplantation? The reason is that this book is concerned almost entirely with nuclear transfer in *amphibians*. Nuclear transfer (grafting) in amoebae preceded that of anurans and urodeles and, as such, it paved the way for the amphibian workers. The trail was blazed not only with concepts and results relating to protoplasmic manipulability but also with systems for contriving microutensils. Concerning the latter, microimplements are no less required for successful handling of amoebae than for wielding amphibian cells. Manufacture of glass micropipettes (Comandon and de Fonbrune, 1934) and microneedles utilizing an early microforge was described (de Fonbrune, 1932). (See Figure A-18 for a modern version of the instrument designed at the Pasteur Institute.) The glass microimplements were then exploited in a series of studies. The first, Comandon and de Fonbrune (1939a) related how they removed the nucleus (*ablation*) of *Amoeba sphaeronucleus* with a fine glass needle. The paper that followed immediately in tandem with the nuclear extirpation report described nuclear transplantation in amoebae (Comandon and de Fonbrune, 1939b). Later, de Fonbrune (1949) illustrated the entire procedure and his photographs are reproduced here (Figure 1-5). In due course, nucleocytoplasmic hybridization was undertaken and the endeavor was successful. In this instance, the nucleocytoplasmic combination was not between species but between ordinary amoebae and amoebae that had been treated with colchicine and which had giant nuclei (Comandon and de Fonbrune, 1942a, 1942b). It may be noted that the nucleus, if it survived the operation, commanded the cytoplasmic volume to assume an appropriate size relative to the inserted nucleus.

The procedure developed by Comandon and de Fonbrune has been utilized by many in recent years (e.g., Muggleton and Danielli, 1968; Goldstein, 1976). A much simplified method for micrury of amoebae was recently innovated (Jeon, 1970). The method allows for the positioning of cells on a moist agar surface where they become immobilized after withdrawal of excess medium. The living amoebae can even be left on the moist agar for several days with-

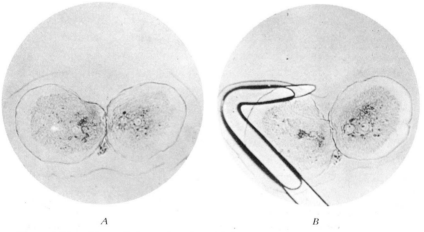

A   B

Figure 1-5A and B. Technique of nuclear ablation and nuclear grafting in amoebae (A. sphaeronucleus). (A) Two amoebae are joined together in a small drop of water. (B) The recipient amoeba is restrained by a glass hook at the left while a fine glass microneedle ("shaft") enters the donor cell at the right. (From de Fonbrune, 1949.)

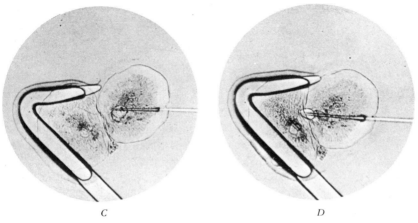

C   *D*

Figure 1-5C and D. Technique of nuclear ablation and nuclear grafting in amoebae (*A. sphaeronucleus*). (*C*) The microneedle, while passing through the cytoplasm of the donor cell, hooks the donor nucleus and moves it in the direction of the recipient amoeba. (*D*) The plasma membranes of the two amoebae are distorted due to the pushing of the microneedle with its impalled nucleus. (From de Fonbrune, 1949.).

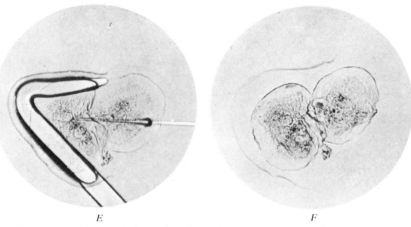

E   *F*

Figure 1-5E and F. Technique of nuclear ablation and nuclear grafting in amoebae (*A. sphaeronucleus*). (*E*) The amoeba on the left is now the recipient of the nucleus of the donor. (*F*) An enucleated amoeba (*right*) and a freshly grafted recipient amoeba (*left*) after removal of the microimplements. Healing of the cytoplasmic wounds occurs rapidly. (From de Fonbrune, 1949.)

out dying. The nuclei are ablated with a microprobe and a new nucleus is pushed in. The donor nucleus can be exposed to air for up to 30 seconds without damage. These are truly hardy nuclei and recipient cells—would that amphibian cells were equally lacking in fragility. Amoeba nuclear grafting was eventually adapted to a number of other organisms (see Appendix IV).

Old Questions Unresolved. The need for a nuclear transplantation procedure, so eloquently expressed by Spemann, Rostand, and others was extended from amoebae to frogs by Briggs and King (1952). Nuclear transfer experiments (discussed in following chapters) are still in progress, a fact that suggests that the problem of the control of gene expression, at least in the cells of higher animals, is far from simple.

It would be nice to conclude this chapter by stating that the old questions posed a half century or more ago are now resolved. They are not. We still do not know how many cell types (if any) in the adult are totipotent. There is controversy concerning the results obtained by nuclear exchange in *Rana* and *Xenopus*. There is good reason to believe that egg cytoplasm has the capacity to order a somatic nucleus to mimic a zygote nucleus. But, we do not know what is in the egg cytoplasm that does the ordering. It is hoped that a reading of subsequent chapters will clarify what we do know concerning cloning experiments, nuclear totipotency, and nucleocytoplasmic interactions as revealed by micromanipulation, and hopefully, will reduce some of the misunderstanding of the results of cloning studies.

CHAPTER 2

Blastula Nuclei Transplanted

The first successful experiments involving the introduction of a living, undamaged embryonic nucleus into an activated and enucleated recipient egg were reported by Robert Briggs and Thomas J. King in 1952. The investigators were at that time associated with The Institute for Cancer Research and Lankenau Hospital Research Institute in Philadelphia. They received the 1972 Charles-Leopold Mayer Prize of the Academie des Sciences, Institut de France, for their pioneering investigations on nuclear transplantation. Briggs and King were the first Americans to receive this, the highest award in biology of the French Academy. The studies of Briggs and King relating to totipotency of blastula nuclei have been extended to later developmental stages in the leopard frog, *Rana pipiens*, and the technique has been adapted to many other amphibian species, insects, and even fish. In retrospect, the frog experiments can be seen to have occurred within a very short time (13 years) after the amoeba nuclear transfer studies of Comandon and de Fonbrune (1939b) and only 14 years after Spemann (1938) expressed the view that nuclear grafting would be "somewhat fantastical." Subsequent chapters of this book will relate how the amphibian cloning technique has been exploited to produce tadpoles whose nuclear donors were derived from such diverse cell types as lens epithelium or renal tumor epithelium, how the procedure may be used to produce polyploid adults, how irradiation of either the nucleus or the cytoplasm separately affect development. However, the present chapter will be concerned with blastula transfer studies.

The 1952 communication from Briggs and King merits careful study. Results are presented which argue convincingly that the method was developed to an extent that it could be used in investigations of nuclear differentiation. It was reasoned that enucleated egg cytoplasm would be a suitable test site for transplanted nuclei because when the cytoplasm is normally nucleated (i.e., fertilized), it develops into all the tissue types characteristic of a species. Yet, egg cytoplasm fails to differentiate any adult cell types at all when totally lacking a nucleus (Fankhauser, 1934; Harvey, 1940) or when lacking a "functional" nucleus (Briggs, Green, and King, 1951). The development of an egg requires a nucleus, and thus, egg cytoplasm is a suitable substrate for nuclear testing of many kinds.

While it is true that cleavage was known to occur in ova deprived of nuclei at the time of the pioneering cloning experiments with *R. pipiens*, two comments should be made at this point. The first involves the nature of nucleusless cleavage. Although it is impressive that eggs lacking chromosomes may furrow, cleave, and produce a "blastula," it should be pointed out that the blastula is not normal. Normal blastulae (Stage 8, Shumway, 1940 – Shumway's illustrations and descriptions of stages of development of *R. pipiens* are reprinted here as Appendix VI) undergo subsequent differentiation and form embryos. Nucleusless blastulae do not. What is perhaps more significant is the fact that nucleusless blastulae are so different from normal blastulae that they can be distinguished from the normal by their appearance when viewed with a dissecting microscope. Normal *R. pipiens* blastulae display a uniformity of cell size within an area, cell diameter grades gradually from the small animal hemisphere cells to the large yolky vegetal hemisphere cells, and the surface has a smooth texture and even pigmentation. These characteristics are altered in a blastula sans nuclei. Thus, one may conclude that a nucleus is a prerequisite not

only for gastrulation but also for normal morphogenesis of a blastula.

The second comment that should be inserted at this point is
that molecular embryology has confirmed the general conclusions
made from observations of blastulae contrived by physical extirpation of chromosomes. Transcription of DNA is blocked by the
antibiotic actinomycin which binds directly to DNA. Actinomycin
treated embryos, while retaining their chromosomes which undergo mitosis, are thus not dissimilar to physically enucleated embryos. Treatment with actinomycin is referred to as chemical enucleation (Davidson, 1976). The developmental fate of an embryo with
a non-transcribed genome is identical with embryos bereft of nuclei, that is, development stops with cleavage. Sea urchin embryos
do not gastrulate and, at most, form abnormal blastulae when
treated with actinomycin (Gross and Cousineau, 1964). Gastrulation is inhibited in amphibians when transcription of DNA is
hindered with actinomycin (Brachet, Denis, and de Vitry, 1964;
Wallace and Elsdale, 1963). Physical extirpation and chemical enucleation of the genetic apparatus both attest to the critical need
for a genome for differentiation to occur. Thus, enucleated egg cytoplasm is indeed a suitable testing site for embryonic nuclei.

Blastula nuclei were scrutinized first because they were generally held to be undifferentiated and undetermined and accordingly
should possess a full spectrum of tissue-type potentialities. A half-
century of blastomere separation experiments and the delayed nucleation studies of Loeb and Spemann suggested (Chapter 1) that
blastula nuclei would prove to be totipotent. (The indications are
particularly compelling in retrospect.) If an egg with a transplanted blastula nucleus could be reared as a normal embryo, and if the
embryo survived and grew to become an adult, the experiment
would attest that the manipulations attendant to transplantation
did not irreversibly damage either nucleus or cytoplasm. The paper
of Briggs and King (1952) contains convincing evidence that (a)
the recipient egg can be reliably enucleated and (b) the development observed must be attributed to the inserted nucleus. The
road paved by Comandon and de Fonbrune (1939b) was now

opened for traffic utilizing amphibian material—which had so long served embryologists well (see tribute to amphibians, p. 173).

Efficacy of Enucleation Technique

It is utterly obligatory that maternal chromosomes be removed in eggs used as cytoplasmic hosts in cloning experiments. This indispensable need for successful enucleation derives from the fact that amphibian eggs, containing maternal genetic material only, are competent to develop remarkably well in the complete absence of paternal chromosomes.

Parthenogenesis, development of virgin eggs without benefit of sperm, has been studied in many species for many years (Wilson, 1901). The phenomenon occurs as a result of laboratory manipulation (Sachs and Anderson, 1970) or occurs spontaneously in nature (Olsen, 1969; Stevens and Varnum, 1974). The early experiments with sea urchin eggs (a French journal once referred to the parthenogens of Jacques Loeb as chemical citizens, the sons of "Madame Sea-urchin et Monsieur Chloride de Magnesium") have recently been extended to mammalian eggs (Graham, 1970).

Gynogenetic embryos can be produced experimentally by fertilizing eggs with sperm of the homologous species that has been treated with chemicals or radiation. The treatment damages the sperm nucleus but has little effect on sperm motility or activating capacity (Moore, 1955). Sperm of a heterologous species, that is, sperm so distant in phylogenetic relationship that its chromosomes cannot replicate adequately in the foreign cytoplasm, will produce gynogenetic development of eggs when used for fertilization. Such a combination is the egg of *R. pipiens* and the sperm of the spadefoot toad, *Scapiopus holbrookii*. Although the toad sperm activates the frog egg, development is exclusively determined by maternal chromosomes and a gynogenetic haploid embryo ensues (Ting, 1951).

Haploid development is almost invariably abnormal with a developmental block occurring at the tadpole stage but development of a few haploid frogs (*R. nigromaculata*) through metamorphosis has been reported (Miyada, 1960). Metamorphosis occurs more reliably with diploid complements of exclusively maternal chromosomes. A novel way of producing gynogenetic diploids is as follows. Eggs of *R. pipiens*, inseminated with sperm of *S. holbrookii*, are heat treated to prevent the loss of the second polar body chromosomes. The retained haploid polar body chromosomes fuse with the haploid mature maternal pronucleus (without the participation of spadefoot toad chromosomes) resulting in a gynogenetic diploid zygote. A few of the embryos became frogs (Volpe and Dasgupta, 1962). Diploid frogs (produced by conventional parthenogenetic procedure) were reported earlier (Parmenter, 1933; Kawamura, 1939). Activation of the egg (completion of the second meiotic division, extrusion of the second polar body, and rotation of the egg), cleavage, and the formation of an embryo ensue after freshly extruded unfertilized frog eggs are pricked with a glass needle dipped in blood or cell extracts. The cleavage initiation factor, a protein, was recently partially characterized by Fraser (1971).

Surely the experiments with haploid and diploid embryos and frogs that develop *without benefit of sperm* argue convincingly that maternal chromosomes have the capacity for extensive and sometimes complete development. Parthenogenesis is indeed a general phenomenon that occurs among both invertebrates and vertebrates, and its occurrence is a very real possibility when any broken cells are injected into eggs that contain the maternal pronucleus. For nuclear transplantation experiments to be meaningful, therefore, the experimenter must be unreservedly assured that the egg chromosomes have been eliminated. This requirement cannot be emphasized too much.

Briggs and King presented persuasive evidence from several kinds of experiments, described below, that maternal chromosomes were eliminated in their nuclear transplantation experiments (Briggs and King, 1952). They activated eggs of *R. pipiens* by pricking with the tip of a fine glass needle. Shortly after activation, "black dots" (i.e., pigment granules clustered about the second maturation meiotic spindle; see Briggs and King, 1953, figure 4, and Subtelny and Bradt, 1961, figure 4) appear followed in several hours by surface puckering and abortive or irregular cleavage furrows. No genuine cleavage or blastula formation occurs, but the surface changes and furrowing are a sign of an activated egg containing a nucleus (parthenogenesis requires a "second factor," a protein, for development to occur beyond simple activation Fraser, 1971). Activated and *enucleated* eggs (accomplished by lifting the spindle, its chromosomes, and a portion of cytoplasm away from the surface of the egg with a glass needle, Porter, 1939 and Appendix II) did *not* undergo puckering and furrow formation within the first several hours indicating success in the enucleation procedure in 631 of 638 operations (99% success).

Cytological evidence for enucleation was difficult to obtain. Twenty-two operated eggs were fixed a number of hours after transplantation. In 21 cases, the exudate from the egg could be seen. Nine of these contained the egg nucleus. Six contained Fuelgen-positive material not definitely recognized as the egg nucleus. No nucleus or Fuelgen-positive material was found in six cases. The cytological examination was useful in that it revealed that enucleation was successful in no less than 68% of the eggs studied. Perhaps there would have been a greater chance of success in finding the egg chromosomes or their remnants if a search were made shortly after enucleation (instead of waiting as in this study until the operated egg had developed to a morula or a later embryonic stage) (Briggs and King, 1952).

A further verification of the precision of the enucleation procedure utilized inseminated eggs. Ablation of the maternal genetic material after fertilization will result in androgenetic haploid embryos. Diploid development is a consequence of failure to remove the egg meiotic spindle and its chromosomes. Androgenetic haploid embryos (i.e., embryos that derive all of their chromosomes from the sperm donor) develop identically to gynogenetic haploid embryos. Most of them will hatch from their jelly membranes and develop to a stage when tail epidermis can be utilized for chromosome analysis. Briggs and King (1952) reported obtaining 337 embryos from 358 operations; *all were haploid*. Thus, haploidy indicated complete success with the enucleation procedure. The difference between 337 and 358 is 21 and that number represented the eggs that did not form embryos. The 21 do *not* represent a failure of the technique. Presumably all 21 would have been haploid had they developed.

Parthenogenetic development stimulated by cytoplasmic granules (a "second factor" substance of parthenogenesis) injection failed entirely among eggs thought to be enucleated. This was in contrast to a low frequency of genuine cleavage and embryo formation obtained after granule injection when the egg nucleus was retained in the host egg (Briggs and King, 1953).

It is not difficult to agree with Briggs and King that "eggs which are pricked and then enucleated will practically never retain the egg nucleus by mistake, and will never develop" (Briggs and King, 1952). The prerequisite of a nucleusless egg to serve as a cytoplasmic host was adequately met with the procedures used by Briggs and King. More recently, it has been shown that ultraviolet irradiation to *Xenopus* (Gordon, 1960a) and urodele eggs and microbeam ruby laser irradiation to *Rana* eggs will serve as well as manual enucleation with a glass needle (Ellinger, King, and McKinnell, 1975). The newer procedures are described in Appendix II. The crucial concern, however, is not the method for enucleation but its effectiveness.

Evidence Revealing That Ensuing Development
Is Attributable to the Transplanted Nucleus

Activated and enucleated eggs, which are competent to undergo no further development without a nucleus, will in a large number of cases (104 of 197 in Briggs and King, 1952) respond to the insertion of a blastula nucleus by cleaving. Further, of the complete blastulae, Briggs and King reported that 74% gastrulated normally and formed embryos. None of the 35 embryos in this pioneering study were haploid. Diploid, and some polyploid embryos, were expected and they were obtained. Polyploidy has since been reported in cloning experiments by a number of investigators and it is reasonably explained as resulting from karyokinesis with an accompanying delay in cytokinesis. Daughter nuclei (diploid) formed from the mitosis of the inserted nucleus fuse in the egg cytoplasm that has not yet cleaved with a consequent formation of tetraploid embryos. Cleavage follows after an initial delay of one cleavage interval. Octaploidy is observed after a delay of two cleavage intervals.

A further proof that the development observed is ascribable to

the implanted nucleus concerns the genetic type of nucleus transferred. Two kinds of experiments have been performed; one involves transferring nuclei from lethal blastulae and the other pertains to use of blastulae cells of mutant frogs (discussed in the
next section).

Egg and sperm hybrids of the northern leopard frog, *R. pipiens*,
and the bullfrog, *R. catesbeiana*, develop to the late blastula or
early gastrula stage and abruptly stop (Moore, 1941). King and
Briggs (1953) reported transplanting androgenetic haploid hybrid
and diploid hybrid blastula nuclei produced by inseminating eggs
of *R. pipiens* with sperm of *R. catesbeiana*. The haploids ensued
when the inseminated ova were enucleated (i.e., when the maternal chromosomes were removed). In contrast to the substantial
number of eggs transplanted with nuclei of the homologous species that developed to larval stages, all eggs that were the recipient
of hybrid nuclei blocked at either the blastula or gastrula stage.
The extent of development of the transplanted egg was limited by,
and was a characteristic of, the lethal hybrid combination. The
hybrid nuclear transfer results can be interpreted reasonably in
only one way, viz., the development of the experimental eggs was
guided by the introduced nucleus.

Genetic and Cytogenetic Markers Confirming
Nuclear Activity in Cloned Frogs

There are in some populations of leopard frogs in Minnesota
and contiguous states individuals with reduced dorsal spotting,
known as Burnsi frogs, and others with a mottled pigment pattern, known as Kandiyohi frogs (McKinnell and Dapkus, 1973; see
also Appendix I). Burnsi and Kandiyohi are both dominant gene
mutations of the northern leopard frog. Blastula nuclei obtained
from Burnsi or Kandiyohi crosses are not only compatible with
enucleated eggs obtained from wild-type *R. pipiens*, but the mutant phenotypes are expressed in the metamorphosed cloned frog
(Figure 2-1) (McKinnell, 1960, 1962b, 1964; Simpson and McKinnell, 1964). Expression of the mutant phenotype in diploid nuclear transplant frogs that develop from eggs of wild-type (spotted)
frogs implanted with diploid mutant (mottled or spotless) nuclei
constitutes substantiation that an inadvertently retained maternal

Figure 2-1. Juvenile frogs produced by the transplantation of blastula nuclei. Both frogs were produced from eggs obtained from wild-type (spotted) Vermont frogs. The diploid frog on the left resulted from the insertion of a Kandiyohi blastula nucleus into the enucleated recipient egg. The frog on the right was produced by transferring a nucleus from a wild-type blastula. The mottled appearance of the frog on the left attests to the fact that development is guided by the inserted nucleus and not by inadvertently retained maternal chromosomes.

nucleus does not promote development. The dominant mottled or spotless genes associated with the introduced nucleus could not have been obtained from the egg of a homozygous recessive female. The mutant phenotype is witness to the activity of the donor nucleus. How is it that we know that the maternal genes are not active? To know this, it is essential that chromosome number be monitored carefully. A triploid frog with the mutant phenotype could develop by the fusion of the maternal pronucleus with the introduced diploid nucleus carrying the mutant tag. Assurance of the lack of maternal chromosomes participating in development is thus dependent upon diploidy of the nuclear transplant embryo and the appearance of the mutant phenotype. The reciprocal experiment, wherein an egg from a mutant frog is host for a homozygous recessive nucleus, serves less well as an indicator of maternal genes not participating in development. Half the mature eggs of a Burnsi frog are expected to be wild-type because adult Burnsi are invariably heterozygous (Moore, 1942). Therefore, spotted progeny from a wild-type nucleus inserted into an enucleated egg of a Burnsi frog does not exclude participation of the maternal nucleus in development. Expression of the spotless phenotype in this combination of nucleus and cytoplasm eloquently states that the maternal chromosomes resided in the egg at the time of nuclear transplantation and subsequently participated in development.

Not all nuclear transfer experiments yield embryos which undergo metamorphosis. Burnsi and Kandiyohi markers are not suitable under these circumstances because the mutant genes are expressed first at metamorphosis. Cells with unusual chromosome number have been used effectively for tagging implanted nuclei. In experiments with newts, haploid nuclei were transferred to normally fertilized, non-enucleated eggs. The presence of haploid nuclei in the ensuing haploid-diploid chimera was interpreted as evidence of survival and proliferation of the inserted foreign nuclei (Pantelouris and Jacob, 1958). Triploid nuclei served as cytogenetic markers for cloning experiments in *Pleurodeles* (Gallien, Picheral, and Lacroix, 1963a) and for tumor nuclei in *Rana* (McKinnell, Deggins, and Labat, 1969; see Chapter 4). Triploid tadpoles in these experiments can only reasonably be interpreted as a consequence of replication of the implanted nucleus with no assistance from maternal chromosomes.

A nucleolar marker (Elsdale, Fischberg, and Smith, 1958) serves in experiments with *Xenopus* in much the same way as mutant pigment pattern genes and ploidy do in *Rana*. A pigment mutant has been used as a nuclear marker in urodele nuclear transfer studies (Signoret, Briggs, and Humphrey, 1962).

*The Predicament of Cytoplasm Accompanying
the Grafted Nucleus*

Damage results to amphibian nuclei freed of their own cytoplasm. Nuclei for transplantation purposes are protected from injury by leaving the donor cell cytoplasm in as undisturbed condition as is possible while allowing for disruption of the cell's plasma membrane. The use of a cell's own cytoplasm as a nuclear medium will continue as a matter of practice until some adequate nuclear medium is developed. So far, none is available. A nucleus that is protected by its cytoplasm and an intact plasma membrane is cloistered from the host egg to such an extent that development fails to occur. The grafted whole cell retains its differentiated appearance, i.e., the nucleus retains its normal size and the plasma membrane remains clearly visible (Figure 2-2). The nuclei of burst

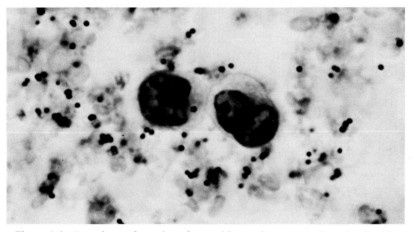

Figure 2-2. Cytoplasm of enucleated egg with two intact transplanted cells. The plasma membranes of the cells effectively isolates their nuclei from host cytoplasm as evidenced by failure of the implanted nuclei to swell or to enter into mitosis. (From King and McKinnell, 1960.)

cells grafted to enucleated cytoplasm provide immediate morphologic evidence that they are interacting with host cytoplasm. Within minutes, the nucleus becomes enlarged and vesicular. What is operationally desirable is an undamaged nucleus that will interact with oöplasm. For the time being, that is a nucleus with a sheath of its own cytoplasm.

The predicament alluded to above is that one seeks with the cloning procedure to characterize the capacity of a *nucleus* to interact with egg cytoplasm but, in fact, one relates what occurs when a *cell* is placed in an enucleated egg. The dilemma provokes two questions relating to the cytoplasm that chaperons grafted nuclei to eggs. One question concerns how much cytoplasm is transferred with the nucleus. The other question, a more important one, asks what effect the escort cytoplasm has on the development of the operated egg.

Some calculations were made concerning volume of donor cytoplasms transferred with the nucleus. Briggs and King (1952) suggested that the small quantity transplanted was not important. They calculated the donor cell volume (nucleus *and* cytoplasm) to be 2×10^{-4} mm^3. If the volume of the egg is 3.4 mm^3, there obtains a ratio of cell to egg of 1:17,000. It was noted however that cytoplasm comprises less than all of the blastula cell. A corrected ratio of 1:20,000 was estimated. Not all cells of an embryo are the same and cell size varies further with the age of the embryo. Hence, several other calculations were subsequently provided. Blastula animal pole cell volume to egg volume was calculated to be 1:48,000 and gastrula animal pole cell to egg was figured to be 1:113,000 (Briggs and King, 1953). Certain donor cytoplasms were estimated to be as small as 1/600,000 the volume of the recipient egg (King and Briggs, 1954).

Regardless of what the volume ratio is, the calculations for the blastula nuclear grafting experiments are probably not particularly important. It is abundantly clear from the frogs obtained by the transfer that many blastula nuclei do indeed have all the potentialities necessary for normal growth and development. Cytoplasm injected sans nucleus does not promote development (of an enucleated egg) and cytoplasm in addition to a blastula nucleus in no way precludes or interferes with normal development. It is possible that introduced cytoplasm from older embryos could affect the potentialities of older implanted nuclei. Nuclei from older donors often result in abnormal embryos upon transplantation. Barth (1964) believes that a consideration of implanted cytoplasm to egg volume ratio may obscure the importance of perinuclear substances. This may be true for older nuclear donors. At any rate, volume ratios in biology are not particularly revealing. Consider the trivial mass of a sperm compared to the enormous bulk of an egg.

What profitably should be considered here is that the need for nuclear protection and the absence of a nuclear medium have resulted in the nuclear transplantation procedure becoming de facto broken cell transfer. Until there is an adequate nuclear medium, there will always be a need for some cytoplasm to be transferred. In a chapter on *blastula* cloning, it may be concluded that the cytoplasm carried across to the host egg does not damage the host egg. Further, there is no way to ascertain if it enhances nuclear performance because of the methodology of the craft.

Cloning and Other Species

The technique of blastula nuclear transplantation has now been extended to a number of other organisms (Appendix IV). The array of species suggests that the procedure is probably adaptable to all amphibians and to other vertebrates including mammals (Bromhall, 1975).

Significance of Blastula Nuclear Transfers

Experiments with blastomere separation and delayed nucleation (Chapter 1) were harbingers of the nuclear transplantation experiments which followed. The pioneers provided experimental evidence relating to the developmental totipotency of certain cleavage nuclei. The results of Briggs and King (1952), as elegant and as exquisite as they were, were not astonishing. This is particularly true, as was stated previously, in retrospect. Perhaps the chief value of the development of a procedure for cloning in frogs was not the results but rather the presentation of a technique that could be exploited further. The results of the exploitations of the technique are presented in the remainder of this book. Although I choose to emphasize the principal value of the nuclear transfer technique as a means of characterization of older nuclei, let it not be forgotten that little was known directly in 1952 concerning totipotency of embryonic nuclei other than the observations of Spemann. Thus, showing totipotency of a nucleus obtained from a blastula containing many thousand cells was a giant step beyond the recognition of totipotency of a nucleus of a 16-cell morula. It should be noted that the frogs resulting from blastula nuclei transplanted to enucleated eggs do not prove that nuclear differentiation had *not* occurred. What the frogs do provide is convincing evidence of the lack of irreversible nuclear change at the blastula stage.

Has nuclear change begun to occur during formation of the blastula? The answer to this query is a probable yes. Consider germ cells if you will (see discussion of Weismann in Chapter 1). Germ plasm is recognized by its staining characteristics during cleavage. Primordial germ cells (PGCs) are cells which contain some of this germ plasm. PGCs can be identified on the floor of the blastocoel (DiBerardino, 1961). Thus, even at the blastula stage, it may be seen that some cells are already morphologically distinct and beginning their unique differentiation. Other cells, those deprived of the germ plasm, are cytologically different and will develop into somatic cells of many types. Does a non-PGC nucleus obtained from a blastula include in its repertoire not only the capacity to form soma but also germ cells? A nucleus obtained from a blastocoel roof (far away from PGCs) was grafted to an enucleated egg. It developed into a sexually mature individual that fathered many offspring (Figure 2-3) (McKinnell, 1962b). Thus, cells beginning to specialize and already segregated as somatic cells retain the capacity to program for functional gametes. The germ cell programming

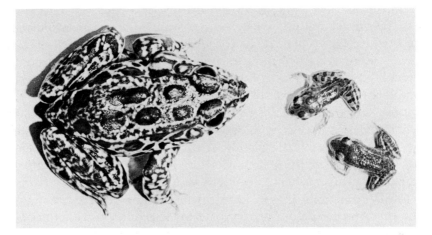

Figure 2-3. Sexually mature nuclear transplant frog which developed from a blastocoel roof nucleus grafted to an enucleated egg. The progeny of the experimental frog demonstrate that non-germinal nuclei can program for normal gametogenesis (McKinnell, 1962b).

is a reversion to a less "differentiated" state of the implanted blastula nucleus.

Many, if not all, blastula nuclei are developmentally totipotent. Is totipotency related to, or due to, a lack of gene activity? We have seen that cleavage, to a large extent, is independent of blastula nuclei or chromosomes. We know this from experiments with sea urchin and amphibian eggs developing in the physical absence of a nucleus or developing after treatment with actinomycin (see discussion at beginning of this chapter). Are blastula nuclei, therefore, biochemically inactive? Indeed not. Transcription occurs, and in amphibian eggs there is a time-dependent change in patterns of RNA synthesis (see Gurdon, 1968c; Humphreys, 1973; Brachet, 1974). Associated with the lack of an ordinary nucleolus, there is little rRNA synthesis until about the onset of gastrulation; tRNA also is delayed until about the same time, but informational RNA is being synthesized throughout cleavage (Brown and Littna, 1964, 1966, Bachvarova and Davidson, 1966; Davidson et al., 1968). New proteins, or an increased rate of synthesis of already present proteins, are detected by electrophoresis during early stages (Malacinski, 1971; Lützeler and Malacinski, 1974). One may ask if any of the "new" proteins are dependent upon nuclear synthesis of their mRNA. This is a reasonable question because it is well known that protein synthesis continues to occur in enucleated embryos. It seems that most proteins that depend upon embryonic genome activity appear after the onset of gastrulation but exceptions to this generalization may occur (see Nemer and Lindsay, 1969; Kedes and Gross, 1969). Thus, "undifferentiated" blastulae nuclei seem to be engaged in activities that zygote nuclei probably are not. These novel genomic activities are not in themselves sufficient to limit in any way the capacity of the blastula genome to be reprogrammed for all kinds of development. Blastula nuclear transfer experiments elegantly demonstrate this fact.

When the nuclear transfer method became available, it was suggested that some early observations be reexamined with the new procedure. Barth (1964) cites Spemann's work with the 1/16th nucleus and delayed nucleation experiment (see Chapter 1). She reminds her readers that this widely cited experiment must be qualified. Development of the operated egg fragment will not proceed beyond the late gastrula or early neurula if the baby hair constriction of the egg is made in the frontal plane and the zygote nucleus is retained in the ventral half. Barth suggests that ventral cytoplasm may be inadequate for complete and normal replication of the zygote nucleus. Note, however, that a ventral embryo will develop if a blastoporal dorsal lip from a normal embryo is grafted to it (Bautzmann, 1927). Whether or not ventral cytoplasm is adequate for normal nuclear replication could be answered relatively easily by nuclear grafting. To learn if a nucleus has been altered by replication in ventral cytoplasm (in contrast to an altered and no longer responsive dorsal cytoplasm), Barth suggests that a ventral 1/16th nucleus be transferred to a freshly activated and enucleated egg. Will normal development ensue? The question asked a decade ago about constricted eggs seems even more relevant today as the importance of cytoplasmic controls of the genome are becoming increasingly studied.

Blastula nuclear transfer reveals a genome with all loci competent to function and with a temporal harmony in gene functioning such that proper sequential activation and repression of genes occurs throughout the period of embryonic development. Transformed juvenile frogs that grow to sexual maturity are attestation to these facts.

It might be added as a postscript to the chapter, with no invidious comparisons intended, that methods such as blastomere separation, constriction of zygotes with baby hair nooses, and cloning have commented cogently and irrefutably concerning genomic equivalence of certain cells. Techniques of molecular biology have yet to do the same.

CHAPTER 3

The Transplantation of Nuclei Obtained from Embryos beyond the Blastula Stage, *Rana pipiens*

Frogs obtained from transplanting animal hemisphere blastula nuclei to activated and enucleated eggs indicated that it was technically feasible to combine parts of cells derived from developmental systems at different stages in ontogeny (i.e., egg cytoplasm at the onset of ontogenetic development could be combined with a blastula nucleus from an embryo that had developed for many hours after fertilization) and to effect the combination in such a manner that irreversible damage to neither the donor nucleus nor the recipient cytoplasm was evident (Chapter 2). The stage was now set to attempt to characterize the developmental potentialities of nuclei obtained from older embryos.

Gastrulation: A Period of Rapid Embryological Change

Rana pipiens eggs are said to be holoblastic. This means that the entire egg mass divides when the zygote nucleus divides. Two blastomeres are formed after the division of the zygote nucleus (Stage 3, Appendix VI—the appendix is material reprinted from Shumway, 1940). The next nuclear division is followed by a cytoplasmic division with four cells formed (Stage 4, Appendix VI). The process of holoblastic cleavage continues rapidly until a hollow ball of cells is formed approximately 16 to 21 hours after insemination at 18°C. The hollow ball of cells (Figure 3-1 *A*) (Stage 8, Appendix VI) is referred to as a blastula.

Gastrulation follows the formation of the blastula. Profound developmental changes occur at gastrulation. The changes at gastrulation may be described at three levels of biological complexity but, obviously, the different modes of description are manifestations of coordinated changes that collectively are termed "gastrulation." The three levels of description alluded to are: (1) changing morphologies, (2) inductive interactions and competence, and (3) changing spectra in the batteries of transcription.

Briefly, the morphological processes may be described as follows: The blastula by a complex process acquires three layers, viz., ectoderm, mesoderm, and endoderm (Figure 3-1). The latent embryological axis becomes apparent such that the future cephalic, caudal, and right and left lateral aspects of the embryo may be identified. Mass migration of surface cells to the interior (a process poorly understood but see review by Johnson, 1974 and discussion of Ballard, 1976) permits the cellular interaction between the interior chordamesoderm (the cellular roof of the archenteron, Figure 3-1 *C* and *D*) with the exterior ectoderm to form the neural plate. The interaction is known as primary induction and has been studied extensively (Spemann, 1938; Toivonen, Tarin, and Saxen, 1976, and Spratt, 1971, chapter 21).

Vital dyes (i.e., stains which tag cells but do not kill them or interfere with their subsequent morphogenetic movements) can be applied to the surface of the embryo so that the fate of cells which move over relatively large distances can be ascertained (Vogt, 1929). Fate maps (diagrams which show ultimate destination and type of differentiation of early embryo cells) can be constructed from the vital dye marking experiments (Figure 3-2).

Competence refers to the capacity of cells or tissues to respond to environmental stimuli by appropriate differentiation. If embryonic ectoderm responds to a neural inductor by forming neural epithelium, it may be said to be competent. The term "determined" applies to a part of an embryo when it is able to develop to its later destiny by self-differentiation, i.e., when its potency is equal to its fate. Prior to determination, potencies (competence to form a diversity of tissue types) exceed fate. Thus, fragments of

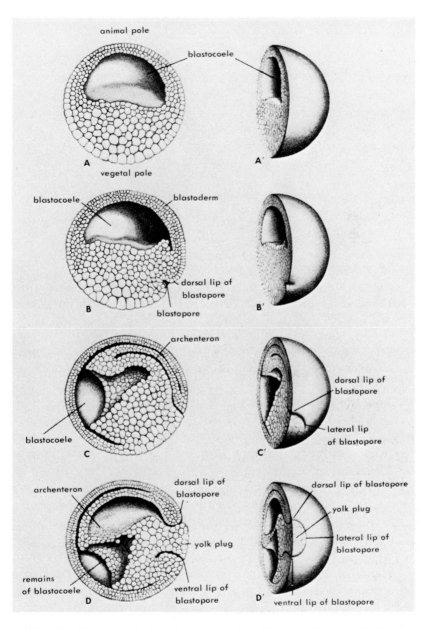

Figure 3-1. Development of the frog embryo from blastula to late gastrula viewed as cut in the median plane (*left*) and from an angle (*right*). *A,A'* blastula (Stage 8^+); *B,B'* beginning of gastrulation with early dorsal lip (Stage 10); *C,C'* mid-gastrula (Stage 11); *D,D'* late gastrula (Stage 12). (The illustration is from Balinsky, 1975, stage numbers are those of Shumway, 1940, reprinted here as Appendix VI.)

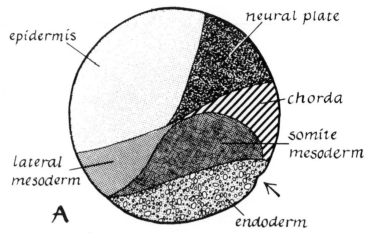

Figure 3-2. Fate map of amphibian blastula derived from vital dye marking experiments. The blastula is viewed from its left side with the site of the future dorsal lip of the blastopore indicated with an arrow. (From *Developmental Biology* by Nelson T. Spratt, Jr. © 1971 by Wadsworth Publishing Company, Inc., Belmont, California 94002. Reprinted by permission of the publisher.)

presumptive epidermis and presumptive neural plate (as identified from fate maps) of a young embryo may be heteroplasticly exchanged between *Triturus cristatus* and *T. taeniatus*. The presumptive neural plate develops in its new location as epidermis and vice versa (Spemann, 1921, 1938). When it is exposed to the environment at the interior of a gastrula, presumptive epidermis of the undetermined embryo develops as part of the notochord, somites, kidney tubules, etc. (Mangold, 1923). After the completion of gastrulation, potency is restricted and becomes equivalent to fate. Late gastrula neuro-ectoderm, while still morphologically identical to skin ectoderm, is limited in its differentiative repertoire to neural structures such as brain or spinal cord depending upon where the gastrula ectoderm is grafted (Spemann, 1919). After determination has occurred, the heart anlage self-differentiates when heterotopically transplanted to the belly of the normal host even to the extent of maintaining the original orientation of the transplant (Copenhaver, 1926).

Weiss (1939) in a description of determination stated: "A cell of a sea urchin blastula has no definitive destination yet. At a later date, however, we find its assignment fixed. The event, or series of events, achieving the change from the indefinite to the definite condition has been called *'determination.'* We can say, therefore, that parts which have started to develop as *equals* become gradually *determined* in various directions, which makes them *intrinsically unequal*, although this inequality, at first, remains latent and in time becomes manifest and discernible. However, the distinction between latent and manifest differentiation cannot be fundamental. For, if two cells which were originally alike become 'determined' to develop in different directions, we must assume that their physical and chemical constitutions have undergone divergent changes, even if there is no external criterion for this. One could almost call *determination* 'invisible differentiation.' "

Davidson (1976) suggested a definition for determination when he alluded to a "battery of early transcriptions." Changing patterns of transcription were described for blastulae in the last chapter and the patterns continue to change throughout gastrulation. The most noticeable change in ribonucleic acid synthesis that occurs at gastrulation is onset of the predominance of ribosomal RNA synthesis (Brown and Littna, 1964; Gurdon and Brown, 1965; Humphreys, 1973) (Figure 3-3).

What does all this mean? We may conclude from the studies of anatomy (morphogenetic movements in the formation of the gastrula), cellular interactions (primary induction and tissues becoming "determined"), and molecular biology (changing patterns of genomic transcriptions) that gastrulation is indeed a time of rapid and impressive change. It is a time when the embryo ceases to be engaged primarily with mitotic fragmentation of the egg mass and gets on with the job of cell specialization. Are the studies of transplanted nuclei derived from gastrulae and later stages in harmony with the observations that tell us that gastrulaton is a time of profound embryological change? Another way of stating this question is to ask if transplanted gastrula nuclei reveal similar profound changes. The narrative below relates that in some ways the answer to these questions is "yes" and in some ways the answer is "no."

Decreasing Yields of Normal Nuclear Transplant Animals Associated with Increasing Age of Donor Nuclei: Technical Damage or Intrinsic Change?

Embryos exhibiting normal differentiation of all three germ lay-

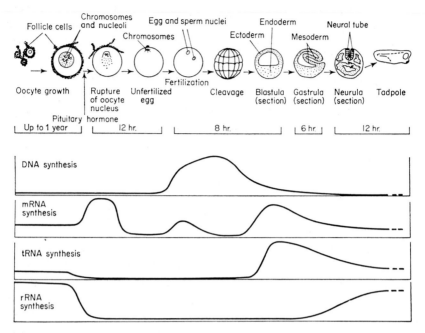

Figure 3-3. Diagram of nucleic acid synthesis related to stages of oogenesis and embryogenesis. The time scale relates more closely to *Xenopus* than to *Rana*. (From Gurdon, 1968c.)

ers develop by transplantation of chordamesoderm (archenteric roof) or presumptive neural plate (dorsal ectoderm) nuclei obtained from late gastrulae (Figure 3-1 *D*; Stage 12, Appendix VI) of the leopard frog. One nuclear transplant animal developed to a stage in metamorphosis. The cells of the roof of the archenteron are determined ("invisibly differentiated") to form mesodermal structures and notochord and the cells of the ectodermal presumptive neural plate are determined to form neural epithelial structures. However, whether one refers to determination, invisible differentiation, or batteries of early transcriptions, the fact remains that developmental properties of late gastrulae have changed notably, but despite those changes, *some* transplanted nuclei manifest substantial developmental capabilities (King and Briggs, 1954).

Perhaps a brief discussion of the significance of the exceptional is in order at this point. As we shall see in much of the remaining material to be discussed, embryologically totipotent nuclei seem to become more and more exceptional as chronological age of the nuclear donor increases. How significant are these increasingly exceptional nuclei? If one were to examine limb regeneration of a previously unscruitinized salamander species and if one were to amputate the right forelimb of 1,000 of these salamanders and if only one of the individuals regenerated a new anterior appendage, I would argue that that one in a thousand salamanders shouts out its most significant biological message. The message, so loud and clear, from the hypothetical exceptional urodele, is that with appropriate conditions (thus far undefined) the form is capable of forelimb regeneration. It would be the next job of the regeneration biologist to define the limiting conditions such that the exceptional becomes the usual.

May I suggest the possibility that the significant observation of nuclear transplantation of postblastular stages is that totipotent nuclei are found in relatively advanced embryos. The fact that the method reveals only a few nuclei that are totipotent should not obscure the possibly immense significance of those few. If the transplanted nuclei whose DNA is capable of total readout with formation of normal progeny are representative of all differentiating (and differentiated) nuclei, then the answer to the question posed above must be "no," i.e., while immense cell and tissue diversifications are occurring, nuclei remain unaltered from that of the zygote.

However, despite the presence of *some* totipotential nuclei from advanced embryos (as revealed by cloning), it is evident that there is a markedly decreased yield of normally developing embryos ensuing from cloning experiments. The decreased yield could be due to one or both of two possibilities. The lower number of viable nuclear transplant embryos resulting from the gastrula or older embryo nuclear exchange could be due to an intrinsic change in the developmental potentialities of embryonic nuclei. Specific nucleotide sequences of the DNA from older embryos could be repressed in a manner so that egg cytoplasm would be an inadequate stimulus for their replication. Another form of intrinsic change, which is perhaps more probable, is that wherein genomic controls of more specialized cells respond sequentially in an inappropriate manner.

Thus, while the nucleus might be stimulated to replicate its DNA, transcriptions would be out of chronological sequence making cell division and proper differentiation impossible (experiments with adult nuclei in *Xenopus*, which replicate DNA but do not divide, described in Chapter 6 are not inconsistent with this view).

Whatever the nautre of DNA replication controls in older embryonic nuclei, the intrinsic nuclear changes could manifest themselves as a restriction in the kinds of differentiated cells that many transplanted nuclei are competent to give rise to or as a restriction in the yield of normal embryos upon transplantation. If the restriction in developmental potentialities of the nuclei that fail to promote normal development are in fact intrinsic to the genome, then nuclei of the later stages may be considered to have become differentiated.

While it is not particularly popular to refer to "differentiated" nuclei, the term is not entirely lacking in merit. A nuclear differentiation concept may be entirely in harmony with current ideas concerning the control of gene activity, i.e., nuclear differentiation does not preclude cell differentiation as an epigenetic phenomenon. Few people seriously consider genetic mutation or a change in chromosomal constitution (selective loss or addition to chromosome sets) as a major cause of cell specialization(Gurdon, 1974b; Ford, 1976). Current thought for the most part centers on selective readout or limited programming of a genome that is common to all differentiated cells within an organism. A more or less stable repression of certain portions of the genome would result in a nucleus that was biochemically and embryologically stabilized as a consequence of differentiation, hence, a "differentiated" nucleus.

Alternatively to the idea that "differentiation" is the cause of reduced nuclear transfer success, one could postulate that an older embryo has cells that are smaller, more fragile, and more difficult to dissociate than blastula cells. Fragility could result from the fact that small cells have less protective cytoplasm and are technically more difficult to transplant. Broken plasma membranes are prerequisite for successful nuclear transfer. Less perinuclear material means less cytoplasmic cushioning. The cushioning may protect the delicate nuclear membrane and nuclear contents. Isolated nuclei swell in isotonic electrolyte solutions. Smaller cells confer less cytoplasmic protection against the damage by nuclear swelling. Older cells generally require more exposure to solutions prior to complete dissociation. The increased exposure to non-physiologic electrolytes, mechanical agitation, and/or proteolytic solutions used for nuclear donor cell dissociation may be damaging. It would seem reasonable, therefore, that reduced yields associated with transplanting nuclei from advanced stages could result from nuclear damage. Many experiments from the decade following the initial cloning studies were designed to distinguish between the alternatives of intrinsic genomic change or technical damage. The reader should keep in mind, however, that exceptional nuclei were obtained from older embryos which exhibited neither damage nor intrinsic change when cloned.

The acute decline in transplanted gastrula nuclei sustaining cleavage (15% from early gastrula nuclei in contrast to 34% from blastula nuclei in those early experiments) could be due to the mitotic stage of the transplanted nucleus. One would expect a different proportion of nuclei to be mitotically active at different stages of development. Indeed, Bragg reported mitotic frequency in the toad, *Bufo cognatus*, to be extremely high for early embryos, especially among animal hemisphere cells. He observed that 81.1% of cells in the cap of the animal pole of a crescent-shaped blastopore stage were in mitosis but there was a rapid drop in the mitotic index at the end of gastrulation (Bragg, 1938). Similarly, the mitotic index varied with embryonic stage in the leopard frog but the rates of cells dividing were not nearly so great. Over 10,000 subsurface nuclei from the vicinity of the animal pole were observed. The following data were reported by Briggs and King (1953, p. 496):

	Interphase	Prophase	Metaphase	Anaphase	Telophase
Blastula	88.7	5.5	2.2	2.2	1.5
Early gastrula	92.1	2.5	1.2	1.4	2.7

Since 34% of blastula nuclei (and more in subsequent experiments) were transplantable, it was concluded that many of the successful transplantations must have been of nuclei in interphase. Since an even larger proportion of early gastrula animal pole nuclei were in interphase compared with blastulae, it was concluded that the stage of mitosis of the inserted nucleus cannot be responsible for the reduced yield of successful nuclear transplantations (Briggs and King, 1953, p. 496).

More recently, stages in the cell cycle between cytokinesis have been correlated with successful nuclear transplantation by a student of mine at Minnesota. The cell cycle in blastula animal hemispheres of *Bombina orientalis* is 70 to 75 minutes at room temperature. Although nuclei selected at intervals about midway between cytokinesis resulted in greater yields of normally developing nuclear transplant embryos, nuclei from all portions of the cell cycle between cytokinesis were able to promote normal development (Ellinger, 1976).

The mechanical manipulations of cloning are such that it certainly seems possible that damage to the nucleus could occur at least part of the time. Damaged nuclei were detected, in fact, in uncleaved cytoplasm after a failure in the transplantation operation. This was reported also in *X. laevis* nuclear transplantation experiments (Chapter 5). Pycnotic vacuolated or fragmented nuclei were found in serial sections of eggs that did not develop (Briggs and King, 1953). Nuclear damage probably occurred because the plasma membrane of the transplanted cell must be broken if the nucleus is to interact with the recipient egg cytoplasm. Gastrula nuclei are probably more sensitive to damage than are blastula nuclei because of smaller size and less cytoplasmic protection (see discussion above).

Because the cell must be broken and its nucleus potentially exposed to the electrolyte solution at the time of cloning, the choice of a transplantation medium is important. It seems obvious that exposed nuclei would be dissimilarly sensitive to different media. A test of eight solutions used in the nuclear transfer dish verifies this statement. Niu-Twitty and Holtfreter's solutions (Appendix III) as transfer media resulted in high yields of successful nuclear transplantations as defined by the formation of blastulae. But, Niu-Twitty solution seemed to be superior to Holtfreter's solution, not in the yield of cleaved recipient eggs, which was about the same for both solutions, but Niu-Twitty solution was superior because over twice as many complete embryos were formed from blastulae as when Holtfreter's solution was used (Briggs and King,

1953, p. 500). The damage that electrolyte solutions can inflict on isolated nuclei, i.e., nuclei deprived of their protective cytoplasm, is discussed in the methods part of this book (Appendix II).

The isolation procedure of cells for nuclear donors affects the yield of nuclear transfer embryos. Donor cells were dissociated mechanically with a glass needle in the first Briggs and King cloning experiments. By 1955, a different method for dissociation was reported. Cells were prepared for transplantation by chemical dissociation and the new method appeared to cause less damage. Cells were isolated by placing the donor embryo in 0.5% trypsin made in a buffered calcium- and magnesium-free Niu-Twitty solution (Appendix III). The enzyme solution allowed for ease in separating the donor embryonic germ layers. The appropriate embryonic layer was then dissociated to the level of individual free cells (Figure A-30) in a 5×10^{-4} M ethylenediaminetetraacetic acid (EDTA) preparation also made up in calcium- and magnesium-free Niu-Twitty solution (Appendix III). Single cells were obtained in 5 to 10 minutes in the EDTA solution. The free cells were then placed in regular Niu-Twitty solution and transplanted. King and Briggs (1955) reported:

Early Gastrula (Animal Hemisphere) Transfers

	# Eggs	Blastulae	Larvae
Dissection with glass needles	135	29	7
Dissociation with trypsin and EDTA	71	29	25

Three times as many nuclear transplant larvae developed from about half as many operations when the donor cells were dissociated chemically rather than mechanically. Different results obtain when nuclei from later stages were similarly treated and compared:

Late Gastrula (Chordamesoderm) Transfers

	# Eggs	Blastulae	Larvae
Dissection with glass needles	242	20 (8%)	4
Dissociation with trypsin and EDTA	83	18 (22%)	2

Successful transplantations (as judged by blastula formation) were increased by almost 300% with chemical dissociation of donor cells. The yield of larvae from the chordamesoderm transfers was not enhanced by chemical dissociation. It was, in fact, low-ered if measured as a yield from blastulae produced by nuclear transfer. King and Briggs conclude that "these experiments indi-cated that the deficiencies in the development of the 'chorda-mesoderm' embryos were not, as we had previously thought, en-tirely the result of nuclear damage but rather reflected an intrin-sic change in the 'differentiative' properties of the chorda-meso-derm nuclei" (King and Briggs, 1955). Contemporary method-ology for cell dissociation suitable for nuclear transfer is discussed in the methods section of this book (Appendix II).

The above experiments suggest that the diminished yield of suc-cessful nuclear transfers may be due to some intrinsic change in the nucleus. If the change *is* intrinsic, several questions can be asked that are subject to experimental testing. How stable is the change? Can the development of a defective nuclear transplant em-bryo be enhanced by parabiosis with a normal embryo? Will nucle-ar fusion of a restricted genome with an undifferentiated genome allow for expression of the repressed or altered gene loci? Are the changes in the nuclei related to the specific cell type of the nuclear donor, i.e., are the abnormalities of endoderm nuclear transfer em-bryos distinct from the abnormalities of embryos produced by transfer of nuclei from another germ layer? Is there a chromosom-al basis for the deficiencies associated with some nuclear trans-plant embryos? These fundamental questions have been investi-gated extensively and are considered in the following paragraphs.

The Stability of Nuclear Change as
Revealed by Isogenic Groups

The progeny of a single endoderm nucleus (or a nucleus of any other type for that matter) can be amplified many times by the procedure of serial nuclear transplantation. Serial transfer is ac-complished by insertion of a nucleus into an enucleated egg in the conventional manner. When the transplanted egg develops to the blastula stage, the transplant blastula is dissociated and its nuclei serve as donors for another nuclear transfer experiment. The pro-cedure is repeated for as many blastula generations as are desired. The resulting embryos all ensue from the original nucleus and are accordingly genetically identical, assuming that somatic mutation and chromosomal damage does not occur.

Many nuclear divisions occur with serial transplantation. It takes about 13 nuclear division generations to form a blastula. If nuclei from a transplant blastula are retransplanted serially for three blastula generations, the donor nucleus will have undergone about 39 (or more) nuclear replication cycles. Since all of the individuals of a nuclear transplant generation have developed from a common blastula that itself was derived from a single nucleus of a cell undergoing specialization, the individuals comprise a clone of genetically identical members. Note here that "clone" is used in the special sense of a group of genetic replicates in contrast to the more general sense used throughout this book, i.e., asexual reproduction (see definition of clone by Kass (1972) in the Preface). Stable nuclear changes, if they occur during gastrula formation, should be revealed in isogenic groups ("clones") of nuclear transfer embryos. Results suggestive of just such a stable nuclear change occurring during gastrulation, a period of great embryological diversification, were reported by King and Briggs (1956).

Gastrula nuclear transfer experiments yield some experimental embryos that arrest at blastula or gastrula stages, others that develop as abnormal post-neurulae, and still others that develop quite normally. Hence, it is possible to envision a gastrula or older embryo as comprised of a mosaic of cells with different developmental capabilities. In a given nuclear transplantation experiment, a small number of presumably random gastrula cells would be chosen for transplantation. If the endoderm of the gastrula is comprised of a mosaic of cells with differing developmental potentialities, it would be expected that cells with differing developmental capabilities would be chosen randomly. However, a blastula that develops from an endoderm nucleus should be made up of developmentally equivalent cells, and serial transfer of the second transfer generation should form an isogenic group of embryos of equivalent developmental potential.

Eighteen nuclear transplant endoderm "clones" were produced as suggested above. "Analysis of the development of 18 clones revealed the following: Whereas test eggs containing different endoderm nuclei developed in the different ways mentioned above, eggs within one clone developed much more uniformly. In some clones all embryos were arrested at gastrula stage, in others they displayed a fairly uniform set of deficiencies in post-gastrula development, and in a few clones almost all embryos developed normally throughout. Furthermore, within any given clone the development in the second and later generations was generally of the same type as that observed in the first generation. In a few clones the deficiencies became more severe in the later generations, but no case of reversal to a more normal type of development was noted" (King and Briggs, 1956).

Not dissimilar observations were recorded with respect to clones of transplant embryos produced from neural cells (DiBerardino and King, 1967). If the variety of deficiencies observed in the first transplant generation were due to a variety of nonspecific injuries associated with the transplantation procedure, then one would expect to observe an equally diverse group of embryos in the subsequent serial transplants. They were not reported. The results with endoderm and neural nuclear isogenic groups argue forcefully that a stable and heritable change in nuclear behavior occurs at gastrulation.

The Stability of Nuclear Change as
Revealed by Parabiosis

The potentialities for forming a specific adult organ or tissue cannot be expressed if an embryo arrests early in development for reasons not associated with formation of that organ or tissue. Latent tissue forming capabilities (i.e., the capacity for programming of nucleotide sequences associated with fully differentiated adult tissue) are therefore not necessarily revealed by the nuclear transplantation technique when the experimental embryo dies prematurely. Information relating to possibly unexpressed potentialities might be obtained by grafting or parabiosing a normal embryo to an experimental embryo destined to undergo an arrest in development. Parabiosis is the laboratory fusion of two separate embryos to form artificial Siamese twins. The normal embryo might provide a physiological milieu that would allow for survival and differentiation of specific adult tissue types that otherwise would be fated to die because of an abnormality of some process that results in death to the experimental embryo.

Despite the possibility of enhanced development, parabiosis of endoderm nuclear transplant embryos to normally fertilized control embryos resulted in no further differentiation of the experimental animals. Thus, a gastrula endoderm nuclear transplant embryo that failed to develop a dorsal blastopore lip at the time when its normal control did, also failed to develop when grafted to a normal host. The normal parabiont developed as a hemi-embryo with the defective nuclear transplant embryo growing as a large mass of undifferentiated cells attached to it. Nuclear transplant embryos with abnormal dorsal lips when parabiosed to normal gastrulae developed as double monsters of varying complexities. Nuclear transplant embryos with normal appearing dorsal blastopore lips frequently develop similarly to control parabionts produced by fusing two normally fertilized gastrulae. It seems clear that parabiosis produced no enhancement of the development of the test embryos, and it was stated that "the nuclear condition responsible for their deficient differentiation is intrinsic in the sense that it cannot be corrected by materials derived from normal parabionts" (Briggs, King, and DiBerardino, 1960).

The parabiosis experiments are of great interest when considered in the context of experiments with grafted tissue from lethal embryos. Haploid embryos, with a single known exception (*R. nigromaculata*, Miyada, 1960), die in early embryonic development. Haploid hybrid embryos (formed with the cytoplasm of one species and the haploid nucleus of another) die even sooner. Is it to be concluded that development stops early because all parts of the genome become inadequate to function at a particular time in development? The answer to this question is "no." When the embryonic tissue that is fated to an early demise is grafted to a normal host, many tissue types develop in the graft that are never expressed in the lethal embryo (Hadorn, 1932; Moore, 1947). Diploid lethal hybrids also persist and differentiate as grafts to normal hosts (Baltzer, 1952). However, as we have seen above, an enhancement of histogenetic potencies does not occur when endoderm nuclear transplant embryos are placed in parabiosis with normal controls.

Since the time of these interesting experiments with parabiosis, much has been learned about tissue immunity and graft rejection. The importance of the thymus to histocompatibility is perhaps no more graphically illustrated than in the mouse mutant *nude*, which suffers from congenital thymic aplasia (Pantelouris, 1968). Grafts of human, cat, chicken, lizard, and tree frog(!) skin are permanently accepted by *nude* mice (Manning, Reed, and Shaffer, 1973). Human tumors previously grown in vitro grow in the mouse mutant (Freedman and Shin, 1974). While there is no *"nude"* frog, thymectomy is now possible in *Rana pipiens* (Turpen, Volpe, and Cohen, 1975). Grafting a fragment of tissue to an athymic host surrounds the graft with diffusible agents of the host and otherwise protects the tissue more than parabiosis, which leaves the graft exposed to a potentially hostile environment. Perhaps endoderm tissue grafted to a thymusless host would have a greater capacity for differentiation than that revealed by parabiosis.

Will Fusion of an Endoderm Nucleus with an Egg Nucleus Result in Enhanced Development?

Diploid embryos that are homozygous for every loci in the entire genome can be produced by inserting haploid blastula nuclei into activated and enucleated eggs. Most recipient eggs cleave on time. They are haploid. Some transplanted eggs cleave after a delay of one mitotic interval. The latter embryos are homozygous diploids. They result from the fusion of daughter haploid nuclei in the egg cytoplasm, which has sustained a delay in cell division. Homozygous diploid embryos develop somewhat better than haploid embryos, but like the haploids, are fated to die before metamorphosis (Subtelny, 1958).

When diploid blastula nuclei are transferred into mature nucleated (haploid) eggs, nuclear fusion with its resultant triploidy occurs (Subtelny and Bradt, 1960, 1963; McKinnell, 1964). A zygote nucleus is formed by the union of a gamete haploid nucleus and a diploid somatic nucleus. Nuclei from homozygous diploid embryos when fused with haploid nuclei from activated (but not enucleated) eggs develop normally and the resulting embryos metamorphose as triploid frogs. Thus, one type of diploid nuclei (the homozygous diploid) fated to undergo an early arrest is enhanced in its developmental capacity when fused to egg chromosomes. The maternal genome presumably introduces needed dominant genes to mask lethal recessives that are expressed in the homozygous diploid embryos (Subtelny, unpublished but cited in Subtelny, 1965b).

Experiments in nuclear fusion were designed to reveal if diploid nuclei obtained from endoderm cells, many of which seem to be restricted in the types of development that they can promote, will develop better if fused to an undifferentiated haploid egg nucleus. Unlike the homozygous diploid nuclei, the endoderm deficiencies behave as if they were dominant and only limited improvement of development follows after fusion with an unrestricted egg nucleus.

As in the experiments of King and Briggs (1956), several kinds of development were observed. Some experiments with first generation serial nuclear transplants of endoderm nuclei (i.e., embryos formed by transplanting nuclei from a blastula which itself had descended from an older endoderm nucleus) produced isogenic groups which uniformly arrested in early development, whether or not the nuclei were transplanted into enucleated or nucleated recipient eggs. Similarly, other first generation serial nuclear transplants of endoderm nuclei developed rather uniformly as abnormal post-neurulae in both the diploid and triploid groups. Finally, normal embryos were obtained in one group of 2n and 3n endodermal nuclear transplants. Thus, fusion with an unrestricted egg nucleus has little effect on the developmental capabilities of an endoderm nucleus (Subtelny, 1965a, 1965b).

Is There Morphologic Specificity of Transplant
Embryos Related to Donor Nuclear Type?

Late gastrula endoderm nuclei when transplanted promote the formation of some normal larvae. About 20% of the complete blastulae derived from endoderm nuclei can be considered to be undifferentiated because they are competent to program for normal development. Some (about 27%) gastrula endoderm nuclei when transplanted guide development only to a blastula or gastrula stage and then the embryo arrests. More than half (53%) of gastrula endoderm nuclear transplantations develop normally through gastrulation but subsequently show abnormalities. The abnormalities are reported to consist primarily of deficiencies in ectodermal derivatives. Endodermal structures appear normal. The collection of specific deficiencies in the endoderm nuclear transplant embryos has been referred to as the *endoderm syndrome*.

The endoderm syndrome of 12 embryos was described in detail by Briggs and King (1957). Epidermis was abnormal in all animals. It was thin or absent in some areas. Pycnotic nuclei and an absence of mitotic activity also characterized the epidermis. The central nervous system was abnormal in all cases. Absence of, or poorly organized, brain tissue and poorly differentiated spinal cord were among the anomalies of the nervous system. Ectodermal derivatives like nasal pits and ears were abnormal or missing in many of the embryos. Pycnotic nuclei were found in sense organs when the latter were present. Neural crest derivatives were poorly developed or entirely lacking in endoderm embryos. Organs of mesodermal origin, in contrast to organs of ectodermal origin, showed substantially fewer abnormalities. Early developing structures were less deviant from the normal than those structures which appear later in ontogeny. Thus, notochord and somites were more normal than later appearing pronephros and blood. Most embryos lacked a heart and none had blood cells. In marked contrast, endoderm tissue appeared to be normal in all embryos. Histological studies of embryos afflicted with the endoderm syndrome would suggest that as differentiation proceeds in the donor embryo, there is loss of the ability of some of its nuclei (as assessed by the cloning procedure) to promote normal development of tissues of other than the donor type.

Following the 1957 study of Briggs and King, it was found that embryos which develop beyond gastrulation and then become abnormal may be divided into two types. The majority of the abnormal embryos are characterized by the description above. However, about one out of five fails to exhibit any consistent pattern of deficiencies. They do not conform to the endoderm syndrome but may have anomalies in any or all germ layer derivatives (Briggs, King, and DiBerardino, 1960).

Endoderm nuclear transplant embryos are not unique in having germ layer specific deficiencies. Karyotype was correlated with an analysis of developmental deficiencies in 47 abnormal *neural* nuclear transplant embryos (DiBerardino and King, 1967). Donor nuclei were obtained from the presumptive neural area of late gastrulae and the neural plate of early and mid-neurulae. The neural nuclear transfer embryos displayed deficiencies in various combinations of germ layers and were associated with abnormal karyotypes or aneuploidy with but one exception. That exception was a group of abnormal embryos which were mainly deficient in organs of mesoderm and endoderm origin. This group of embryos was the only group in which all embryos possessed the normal karyotype. The neural nuclear transplant embryos of normal karyotype were reported to have extensive differentiation of ectodermal tissue types but notochord and somites appeared to be retarded and contained pycnotic nuclei. Heart and pronephros were also improperly formed. Gut and its derivatives were poorly differentiated. Cytological examination of the liver revealed that about half of its nuclei were abnormal. The esophagus was either completely lacking or was not patent. It would thus seem that neural nuclear transplant embryos displayed abnormalities "distinct from those found previously in endoderm nuclear transplants. The pattern of deficiencies is consistent with the neural origin of their donor nuclei" (DiBerardino and King, 1967).

The unique contribution concerning embryos cloned from neural nuclei adds considerable credence to the many studies of endoderm nuclear transplant embryos—especially with reference to the specificity of the kinds of abnormalities observed in the experimental animals. It is not generally recognized that only endoderm nuclei have thus far been reasonably fully characterized by cloning. That the specificity of deficiencies in endoderm embryos is attributable to nuclear characteristics and is not due to the procedure is strongly suggested by a single clone of neural nuclei. The single group of neural embryos to which I refer is the one that was euploid with a normal karyotype and that displayed deficiencies quite different from endoderm embryos. It seems obvious from consideration of the endoderm- and neural-cloned animals that development of nuclear transplant embryos cannot be considered apart from their chromosome complement. Therefore, chromosomes and their relationship to cloned embryo development will be considered next.

*Chromosomes and the Development of
Cloned Embryos*

Is there a chromosomal basis for the abnormal development following the cloning of many endoderm nuclei? It has been known for three-quarters of a century that aneuploid embryos do not develop normally in contrast to diploid and some polyploid embryos (Boveri, 1902; see also Fankhauser, 1945, 1952). It would be of value, therefore, to know the chromosome constitution of embryos produced by inserting nuclei into enucleate ova. Indeed studies were made of endoderm transplant embryos with reference to their chromosomes not dissimilar to the study of neural transplant embryos described above.

Late gastrula-cloned embryos that arrested early in development were found to be aneuploid. The abnormalities of the embryo were probably caused by unbalanced chromosome constitutions and hence these embryos do not constitute a valid test for nuclear differentiation (the *origin* of the aneuploidy is of course of considerable interest and will be discussed below). Normally developing cloned embryos were observed that were not only euploid but also had a normal karyotype. They were considered to be evidence that some nuclei from late gastrulae were equivalent to the zygote nucleus with respect to developmental potency. However, some euploid embryos developed with the endoderm syndrome. These endoderm-syndrome embryos were the nuclear transplants of interest to embryologists. Here no chromosomal abnormality was obvious but development followed an abnormal but characteristic course. This was perhaps the best evidence that some sort of intrinsic nuclear change had occurred. These intriguing chromosome studies which correlated karyotype with development were reported in Briggs, King, and DiBerardino (1960).

It would seem that normal appearing chromosomes do *not* always promote normal development. There may be some value therefore in distinguishing between *developmental* totipotency and *genetic* equivalence (Stern discussion to King and Briggs in King and Briggs, 1956, p. 290). A normal karyotype associated with limited developmental potential would suggest that the entire genome is present but its capacity to function in egg cytoplasm reveals that at least some genetic loci are not functioning or are functioning in some manner so as to interfere with normal development. It should be noted, however, that conventional karyotypic analysis lacks sufficient resolution to reveal minute chromosomal change. Small deletions or translocations of developmental significance would not necessarily be detectable even by modern banding procedures (Comings and Avelino, 1975). Thus, the normal appearing karyotype associated with endoderm-syndrome embryos may in fact be an abnormal chromosome complement. The nature of the controls that alter developmental totipotency is of course a central problem of developmental biology. Gene activity controls, especially in the context of nuclear transfer, have been considered recently by Gurdon (1974b) and Holliday and Pugh (1975).

Transplanted blastula nuclei generally retain a normal euploid complement of chromosomes. Many gastrula nuclei become aneuploid upon transplantation. Although the extent of embryonic development that aneuploid embryos undergo is not an adequate test of the state of differentiation of the donor nucleus, the fact that the embryo is indeed aneuploid reveals that changes have occurred in the capacity of the nucleus for normal replication in egg cytoplasm. Some of the aneuploidy is doubtlessly due to technical damage to the cloned nucleus but competent nuclear transplanters believe that nuclear damage plays only a minor role in the genesis of aneuploidy. It should be noted here that this discussion concerns unbalanced chromosome sets observed while nuclei replicate in homologous egg cytoplasm. Blastula nuclei often become aneuploid when they replicate in the cytoplasm of a different species. Since the aneuploidy of nucleocytoplasmic hybrids probably arises from different causes, it will not be considered here (see Chapter 7 for a discussion of nucleocytoplasmic hybrids).

Subtelny (1965b) speculated concerning the nature of the changes in nuclei that are involved in the failure of injected embryonic nuclei to maintain a normal karyotype in egg cytoplasm (see also Gallien, Picheral, and Lacroix, 1963b; and Gallien, 1974). He concurs with others (Briggs, Signoret, and Humphrey, 1964) that asynchrony of DNA replication could lead to chromosome abnormalities upon transplantation. It may be of interest to note that differential replication of DNA has been reported in many species, including axolotl embryos (Signoret and Lefresne, 1970), newt spermatocytes (Wimber and Prensky, 1963), *R. pipiens* (Flickinger, Freedman, and Stambrook, 1967), and other higher animals (see Hsu, Schmid, and Stubblefield, 1964).

In the study of Flickinger, Freedman, and Stambrook (1967), regions of late replicating DNA at one or both ends of many metaphase chromosomes are found in late neurulae and tailbud embryos but *not* in cells of blastulae or mid-gastrula stage embryos. The frog labeling experiments are not precisely what would be expected if asynchronous DNA synthesis were indeed the only cause of aneuploidy among the transplant embryos. The lack of late-labeling chromosomes among blastula cells fits with expectations as do late-labeling chromosomes found in neurula and tailbud cells. What fits less well is the absence of late-labeling chromosomes among gastrulae cells—a phenomenon that one might expect if asynchrony were the sole and unique cause for aneuploidy in cloned leopard frogs.

A recent study by DiBerardino and Hoffner (1970) suggests that aneuploidy in nuclear transplant embryos is traceable to the first response of the implanted nucleus to the recipient cytoplasm. Some inserted nuclei fail to enlarge and inadequate chromatin decondensation occurs with a resultant formation of chromatid bridges and subsequent chromatid breakage. Further, many cloned embryos which develop abnormally exhibit severe and dense condensation of chromosomes.

The observations of DiBerardino and Hoffner (1970) simply reveal that gross chromosomal abnormalities detected in abnormally developing embryos may be traced back to within hours of nuclear insertion. While unbalanced chromosome complements may be responsible for abnormal development, a critical question concerns why the chromosomes replicate improperly in the first place. Flickinger, Freedman, and Stambrook (1967) reported that cell cycle length increases markedly from blastula to tailbud. Generation time and G_1 vary most but S and G_2 of the cell cycle vary also with developmental stage. Can cell cycle duration affect transcription (Detlaff, 1964) of inserted nuclei, and because of this, or because of other reasons, affect chromosomes such that they replicate abnormally in egg cytoplasm? Whatever is the answer to this question, the fact remains that developmentally significant reprogramming of differentiating nuclei is contingent upon essentially flawless chromosome replication. Such replication does not always occur in nuclear transfer experiments.

How Late in Development Are Totipotent Nuclei Found?

Cloned endoderm nuclei obtained from late gastrulae will program the kinds of development previously described in this chapter. Although the majority of transplant embryos develop abnormally, a few develop into normal larvae. The cleavage promoting capacity of mid-neurula endoderm nuclei is reduced from that of gastrula endoderm nuclei, and, the majority of transplant embryos arrest as blastulae or gastrulae with the remainder developing abnormally with but one exception. No normal transplant larvae were obtained when the donor nuclei were derived from tailbud endoderm cells. Of nine blastulae obtained from tailbud transfers seven arrested as blastulae or gastrulae and the other two developed abnormally (Briggs and King, 1957).

Primordial germ cell nuclei, derived from a tadpole just prior to the onset of feeding (Stage 25, Appendix VI), contain many totipotent nuclei. Transplanted germ cell nuclei caused cleavage in 43% (of 410 transfers) of the recipient enucleated eggs, and 40% of the completely cleaved eggs developed normally. However, no normal tadpoles were obtained from endoderm (Stage 23, Appendix VI) control nuclear transfer experiments. Control blastula nuclear transfers promoted normal development in the expected high percentage. The germ cells of pre-feeding embryos and the endoderm cells of slightly younger embryos are about the same size, thus "the difference in results between germ cell and endoderm nuclear transfers (40% compared with 0%) would not be explainable on the basis of technical difficulties in carrying out the transfers, but should reflect real differences in the developmental capacity of the nuclei of these two cell types" (Smith, 1965).

Nuclear transplantation of differentiated male germ cells derived from metamorphosed or adult frogs results in a low frequency of cleavage (16% and 10% respectively). Most of the blastulae arrest during gastrulation. None developed into normal larvae. Thus, genetically totipotent spermatogonia seem to be developmentally restricted (DiBerardino, King, and Bohl, 1966). This is in contrast to developmentally totipotent primordial germ cells of the late pre-feeding embryo (Smith, 1965), a condition unpredicted but the subject of a previous speculation (Stern discussion to King and Briggs in King and Briggs, 1956, p. 290).

Studies have been published, and are in progress, of the capacity of neoplastic nuclei derived from adult frogs to guide embryonic development. These studies constitute the subject of Chapter 4. Muggleton-Harris and Pezzella (1972) have transplanted nuclei derived from "aged" lens epithelium. Complete embryonic development was claimed to have occurred in these exceptional experiments. Their studies will be discussed in Chapter 6.

Spermine is an almost universally distributed polycationic amine that perhaps plays a role in regulation of nucleic acid metabolism (Raina and Janne, 1970). Polyamines may be specific activators or inhibitors of enzyme reactions (Raina and Janne, 1975). Spermine can be bound in the groove of double helical DNA and when this happens, the negative charges of DNA are neutralized and the molecule becomes more flexible. Spermine is an essential growth factor for some microorganisms. It prevents swelling of mitochondria and protects some microorganisms from loss of cellular constituents in hypotonic media. Polyamines have been used to protect isolated nuclei. The reason for mentioning the polyamine spermine at this point is that remarkable results have been reported concerning its effects and the effects of low temperature on results of nuclear transfer experiments. Normal larvae were obtained by the transplantation of endoderm nuclei obtained from tailbud donor embryos with spermine in the transfer medium and lowered laboratory temperature (Hennen, 1970). The yield of cloned normal larvae was about 23% of total transfers or 62% (16 of 26) of complete blastulae. *None* would be expected using conventional nuclear transfer procedure. Hennen (1970) stated: "If normal differentiation involves the selective repression of genetic information, then repression, however stable it might be under normal circumstances, is reversible as far as nuclei from tailbud presumptive midgut are concerned."

What is the answer to the question posed in the heading of this section? The answer at the present time is complex (McKinnell, 1972). In the isolated instance of a special cell type of *R. pipiens* treated in an unusual manner, totipotency seems to persist through adulthood (Muggleton-Harris and Pezzella, 1972). Conventional cloning procedure suggests that nuclear changes occur quite early in development (Briggs and King, 1957), changes which are manifested in improper mitotic behavior at the very onset of contact of inserted nucleus with host cytoplasm (DiBerardino and Hoffner, 1970). However, manipulation of the physical and chemical environment of dissociated cells and recipient eggs seems to alter or modify these early nuclear changes (Hennen, 1970). The question will be considered again when the results obtained with *Xenopus* are reviewed (Chapter 5) and when cloning of adult cells is considered in detail (Chapter 6).

Transplanted Cytoplasm Considered Again

It has been asserted that a small quantity of cytoplasm is transferred with the nucleus to the recipient egg in the cloning procedure (Chapter 2 and Appendix II). Juvenile frogs and frogs that develop to sexual maturity attest to the failure of the transferred cytoplasm to have a deleterious effect in cloning of blastula cells. Because the majority of transfer embryos develop abnormally when the donor nucleus is obtained from an older embryo, the biological effect of contaminating cytoplasm is considered again here briefly.

Endoderm cytoplasm was inserted into enucleated and fertilized eggs. Cytoplasm was obtained by partially pulling an endoderm cell into a micropipette and then cutting away the portion of the cell outside the micropipette containing the nucleus. Twelve enucleated eggs were injected with endoderm cytoplasm. None cleaved. Thus, endoderm cytoplasm failed to elicit cleavage in an enucleated egg. Cytoplasm from late gastrulae, mid-neurula, and tailbud embryos was injected into 68 fertilized eggs. The development of injected eggs did not differ from uninjected control eggs. The majority of experimental and control embryos developed normally. A few abnormal embryos occurred in both the injected and control groups. However, the abnormalities were not those of the endoderm syndrome. Briggs and King (1957) stated, "there was thus no evidence that endoderm cytoplasm by itself can either elicit cleavage in enucleated eggs or alter the cleavage or differentiation of normally nucleated eggs."

While it is true that cytoplasm from endoderm cells injected into zygotes does not hamper normal development in most cases, that same kind of cytoplasm in intimate proximity to a transplanted endoderm nucleus may impede the passage of proper regulatory signals from egg cytoplasm to the inserted nucleus. "Regulatory signals" are known by the reprogramming effect of cytoplasm on inserted nuclei. Further, the existence of such signals is inferred from experiments that reveal a bidirectional flow of nonhistone proteins between egg cytoplasm and inserted nuclei. ^{3}H-tryptophane-labeled proteins leave endoderm nuclei inserted into unlabeled mature egg cytoplasm, but proteins labeled with ^{3}H-lysine primarily remain in the nuclei (DiBerardino and Hoffner, 1975). Cold late gastrula endoderm nuclei accumulate nonhistone proteins during the first cell cycle after nuclear transfer (Hoffner and DiBerardino, 1977; see also related experiments of Brothers, 1976). The elucidation of the molecular events by which the cytoplasm effects a reprogramming of inserted nuclei is perhaps the most exciting area in cell biology today.

CHAPTER 4

Transplantation of Nuclei Obtained
from Tumors of *Rana pipiens*

Do cancer cells arise as a result of somatic mutation (Schultz, 1959), or do cancer cells arise by epigenetic means that leave the genome unaltered from other normal adult cells (Braun, 1974)? This is a question that can be asked by a person trained in the cloning art, and partial answer to the question is already available. The rationale is simple. If a cancer cell has an intact genome containing all of the nucleotide sequences for normal development, and if that genome can be provoked to express itself by nuclear transplantation, then the resulting cloned animal would stand in witness to the completeness of the cancer cell genome. Alternatively, deletion or alteration to sequences of purines and pyrimidines that comprise the cancer cell DNA would result in a genome that could never be expected to program for normal development. Thus, cloning might reasonably be expected to provide insight into whether a particular tumor has developed as a consequence of genetic or epigenetic modification.

The question of genetic versus epigenetic causation of cancer seeks insight into one of the most fundamental aspects of neoplasia. If it could be shown that some or many tumors arise as a malfunction of the differentiative process, then there is hope that normalization of the tumor could be effected so that existing reverted tumor cells and their progeny, rather than giving rise to more malignant cells and ultimately causing death, could instead resume tasks useful to the survival of the individual.

Reversible Tumors

If cancer cells of some types arise through epigenetic mechanisms that do not alter the structural integrity of the DNA, there is at least a possibility that the process may be reversible. Reversion should not only occur in the laboratory as a result of manipulation designed to return a malignant cell to normality but it should also occur "spontaneously" if natural circumstances obtain which are permissive and if the potentiality for normality is present.

A review of the literature of tumor cell "normalization" is rewarding because it reveals the diversity of tumor types that have been manipulated to become more normal and because it demonstrates that a variety of technical maneuvers may be exploited to force or to influence neoplastic cells to express their intrinsic capacity for normal differentiation. It is intellectually gratifying to ponder the recent developments concerning the controlled reprogramming of tumor cells because they provide compelling arguments that the advances in genetic and molecular biology during the decade of the sixties will be valuable in understanding the neoplastic state.

Amphibians are easy to rear and have enormous cells and chromosomes. Accordingly, it is not surprising that there was a report a quarter of a century ago that frog renal tumor cells could, under the influence of a regenerating "field," give rise to normal muscle, cartilage, and connective tissue cells (Rose and Wallingford, 1948; Rose, 1949). (Note, however, that another study was not successful in repeating these observations, Ruben, 1956.)

Pierce and his associates reported conversion of malignant cells to a multiplicity of somatic cells in mouse teratocarcinoma and they have observed that rat squamous cell carcinoma gives rise by mitosis to normal cells (Pierce, 1972, 1974, 1976). Neural tissue, cartilage, striated muscle, etc., have been obtained from mouse testicular teratoma in a similar study by Stevens (1960). Normal cell progeny, as well as tumors, ensued when small clumps of teratocarcinoma were injected into early mouse embryos (Papaioannou et al., 1976). Progeny of single teratocarcinoma cells, injected into mouse blastocysts, contribute to many types of normal somatic tissues (Illmensee and Mintz, 1976). Those who wish to consider further the phenomenon of highly malignant cells expressing more benign characteristics may read a number of articles and reviews (Bloch-Shtacher, Rabinowitz, and Sachs, 1972; Braun, 1974; Markert, 1968; Meins, 1974; Pierce, 1972, 1974, 1976). Epigenetic processes are considered in a volume concerning cancer and cell differentiation (Sherbet, 1974).

Reprogramming a Tumor Cell Genome and Developmental Biology

Developmental biologists should be as interested in tumor nuclear transfer studies as pathologists. Tumor cells in an adult organism are descended from fully differentiated adult cells (unless, as seems unlikely, they are progeny of embryonic rests). As descendants of differentiated cells, any capacity of their nuclei to promote differentiation must reflect the potentialities of the normal cell of origin. Thus, genetic information resident in a tumor nucleus is unlikely to be greater than the genetic information present in the normal nucleus from which it was derived. Further, if all somatic nuclei of a species are constructed of qualitatively identical DNA and tumor cells and their normal cells of origin differ in only epigenetic aspects, then it would seem that the two cell types would afford elegant material for the study of gene repression and gene activation control mechanisms.

Although nuclear transplantation procedures are well developed in the anuran *Xenopus* (Chapter 5) and tumors of the species have been described (Balls and Clothier, 1974), similar experiments in the African clawed toad have not been reported. Nuclear transfer is also well known for the urodeles (Chapter 5) but nuclear transplantation of any of their tumors (Brunst, 1969) has not been reported either.

It seems to me that if one is to understand the complexities of tumor nuclear transplantation, one should know something about the biology of the tumor under consideration. Certain aspects of the etiology, cytology, and epidemiology of the Lucké renal adenocarcinoma will be presented so that the nuclear transplantation studies discussed later in this chapter will be more meaningful.

The Renal Tumor of Leopard Frogs: Distribution and Prevalence

A kidney cancer of the northern leopard frog, *R. pipiens*, was described by Lucké (1934). Frogs with the renal tumor were thought to be limited in their distribution primarily to northern Vermont and contiguous areas of Quebec Province, Canada, but examination of frogs obtained from midwestern dealers (McKinnell, 1965, 1969, 1973a) indicates that susceptible populations of frogs extend to Minnesota (at least 1,000 miles to the west of Vermont). Leopard frogs obtained from North Dakota, Louisiana, and more western states may be free of spontaneous tumors (McKinnell, 1969; McKinnell and Duplantier, 1970).

Prevalence of tumors detectable at autopsy (Figure 4-1) varies seasonally with neoplasms being scarce during the warm months and relatively abundant during winter and early spring (McKinnell and McKinnell, 1968). I reported a tumor prevalence of 10.5% among frogs collected prior to emergence from winter hibernation from a lake in a west central Minnesota county (McKinnell, 1969, 1973a).

Herpesvirus Etiology of the Frog Cancer

Herpesviruses are associated with several "natural" cancers of animals such as frogs, chickens, guinea pigs, and rabbits (Goodheart, 1970). Further, several human cancers are linked with herpesviruses. These include Burkitt's lymphoma (Burkitt, 1972; Epstein, 1972), nasopharyngeal carcinoma (Shanmugaratnam, 1972; de-Thé, 1972), Hodgkin's disease (Stepina et al, 1976), and cervical carcinoma (Sabin and Tarro, 1973; Kaufman and Rawls, 1974). It may be of interest to some, especially those with an affection for frogs, that the Lucké renal adenocarcinoma was the first cancer thought to have a herpesvirus etiology.

How do we know that the frog tumor virus is a herpesvirus? Herpesviruses are DNA viruses. It has been known for over a decade that the frog renal tumor virus is a DNA virus. Viral nucle-

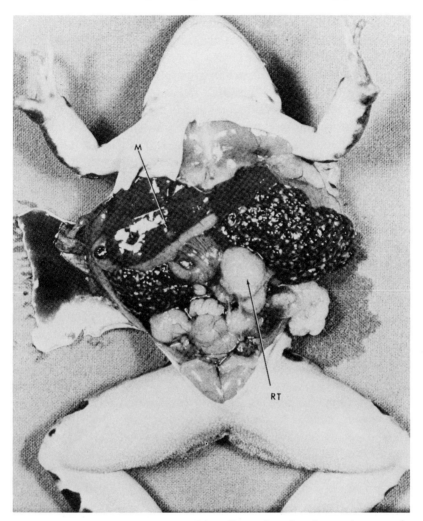

Figure 4-1. Sexually mature *Rana pipiens* dissected to show large primary renal adenocarcinoma (*RT*) with several metastatic masses in the liver (*M*). (From McKinnell, Steven, and Labat, 1976.)

oids in frog tumor thin sections are unaffected by trypsin and RNase but are completely digested by DNase (Zambernard and Vatter, 1966). Negatively stained preparations of the frog renal tumor virus reveal that it has icosahedral symmetry with a capsid consisting of 162 capsomeres (Lunger, 1964) identical with other herpesviruses (Wildy, Russell, and Horne, 1960). Thus, there is no doubt that the Lucké tumor virus is a herpesvirus.

The frog tumor virus occurs in great abundance in spontaneous tumors. Tritium-labeled thymidine radioautographs of tumor cells with mature virions contain silver grains over the virus core, which provides confirming evidence that the viral nucleic acid is DNA (Stackpole, 1969). The abundant virus, which has been estimated to occur with a frequency of 5×10^{11} virus particles per gram of winter tumor (Toplin, Brandt, and Sottong, 1969), permits the characterization of its DNA and the comparison of its DNA with that of other herpesviruses (Wagner et al., 1970).

Electron microscope studies of frog renal tumors reveal virus particles in some tumors (Fawcett, 1956) or in all (Zambernard, Vatter, and McKinnell, 1966; McKinnell and Zambernard, 1968). The presence or absence of formed virus particles is believed to depend primarily upon temperature (Zambernard and Vatter, 1966; Zambernard and McKinnell, 1969; Stackpole, 1969). Fine structure studies made of field-collected frog tumors reveal that tumors of frogs collected prior to the onset of hibernation are devoid of virus particles detectable with the electron microscope (Figure 4-2). Virus particles are abundant within a few days after the beginning of hibernation (Figure 4-3) (McKinnell et al., 1972) and persist for as long as 52 days after the host frog leaves hibernation with the return of warm weather (Figure 4-4) (McKinnell and Ellis, 1972a, 1972b). It thus becomes clear with a perspective of time that the virus-tumor relationship, previously thought to be an uncertain one, is in fact quite predictable. Tumors of unmolested frogs from nature during the cold period of the year will invariably contain herpesviruses.

What happens to viruses in tumors during the warm period of the year? There is now ample evidence for the presence of the viral genome, in a latent state, in "virus-free" renal tumors. Extirpated chunks of tumors containing no detectable virus particles will form herpesviruses in vitro when chilled (Morek, 1972). "Virus-free" tumor transplanted to the anterior eye chamber (Mizell, Stackpole, and Halpern, 1968) of tumor-free adult frog hosts similarly are provoked to produce herpesviruses when chilled. Lucké

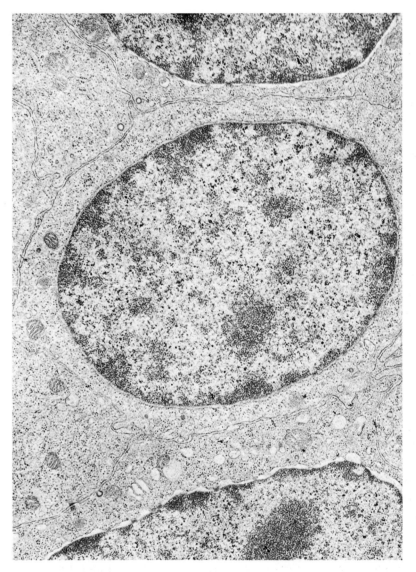

Figure 4-2. "Virus-free" renal tumor obtained from a prehibernating frog collected by the author near Diamond Lake, Kandiyohi County, Minnesota. × 13,200.
(From McKinnell, 1973a.)

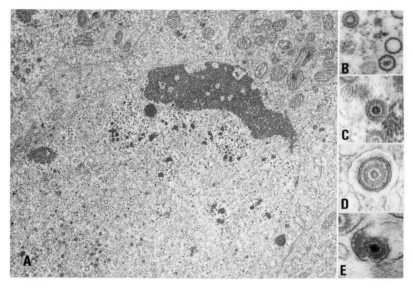

Figure 4-3. (*A*) Electron microscope study of a spontaneous renal tumor cell containing herpesviruses (× 8,200). The tumor-bearing frog was collected by the author in Kandiyohi County, Minnesota. Ontogenetic stages of an oncogenic virus are shown to the right (× 45,600). (*B*) Three types of nuclear virus particles — empty capsid, capsid with double-walled core, and nucleocapsid with dense core. (*C*) Cytoplasmic virus with dense coating. (*D*) Cytoplasmic virus with envelope. (*E*) Extracellular-enveloped virus. (From McKinnell et al., 1972.)

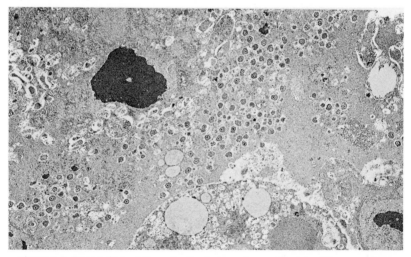

Figure 4-4. Mature herpesviruses in the lumen of a renal tumor. The tumor-bearing frog was captured June 3, 1971, in rural Minnesota. Emergence from hibernation and spawning occur in mid-April. (From McKinnell and Ellis, 1972b.)

tumor herpesvirus transcripts are detected in tumors bereft of detectable virions (Collard et al., 1973). Thus, even those tumors which do not contain frank herpesviruses can be induced to produce them or viral DNA-specified messenger RNA can be detected in them. Koch's first postulate (the recognition that an etiological agent causes a specific disease) requires that the agent be present in all instances of the disease. Clearly, this first postulate has been met in the Lucké renal adenocarcinoma of the northern leopard frog.

How does one isolate viruses and grow them in pure culture (the second postulate of Koch)? This of course is not possible with viruses and the difficulty of strictly applying Koch's postulates to viral diseases has long been recognized (Rivers, 1936). Nevertheless, viruses can be obtained by centrifugation of tumor homogenates and these viral preparations cause the formation of tumors when injected into immature (susceptible) animals, i.e., pre-feeding tadpoles (Tweedell, 1967, 1969; Tweedell and Wong, 1974; McKinnell and Tweedell, 1970). The presumed etiological agent can then be recovered from the induced tumor by cold-treatment, thereby fulfilling the last of Koch's postulates (Naegele, Granoff, and Darlington, 1974). One must now agree, without any reservation, with the cell biology correspondent of *Nature* who wrote: "The Lucké adenocarcinoma still offers what seems to be the best opportunity to test the oncogenic potential of a herpes virus" (Anonymous, 1971; McKinnell, 1973a).

Frog Tumor Chromosomes

Normal development does not occur in the absence of a balanced set of chromosomes. Not a specific number of chromosomes but a specific combination of chromosomes is a prerequisite for normal development (Boveri, 1902). Although diploid (2n), triploid (3n), and tetraploid (4n) euploid amphibians develop normally beyond metamorphosis (Chapter 7), deviations in chromosome complement (e.g., $2n-1$, $2n-2$, $2n+1$, $2n+2$) that produce unbalanced chromosome combinations (aneuploidy) result in abnormal development (Fankhauser, 1945, 1952; see also discussion in Chapter 3). It would therefore be of little value to attempt to characterize the developmental potentialities of a tumor cell lacking an euploid chromosome complement.

Fortunately, for the purposes of cloning, many cells of the Lucké renal tumor of frogs appear to be euploid. Lucké (1939) reported that abnormal mitoses were uncommon and when the number of chromosomes could be counted with certainty, the diploid number was found. The diploid number of chromosomes for *Rana pipiens* is 26 (DiBerardino, 1962) and this number was reported for cultured renal tumor cells by Duryee and Doherty (1954) and Freed and Cole (1961).

Aceto-orcein chromosome squashes of primary and cultured renal tumors reveal that many tumor cells have a karyotype (Figure 4-5) not distinguishable from normal embryo (Figure 4-6) and normal adult (Figure 4-7) chromosomes. Karyotypic analysis of some tumor chromosome complements disclosed aneuploidy and non-specific structural abnormalities of chromosomes. The variations of the neoplastic chromosomes are similar to abnormalities occasionally observed among chromosomes of normal cells

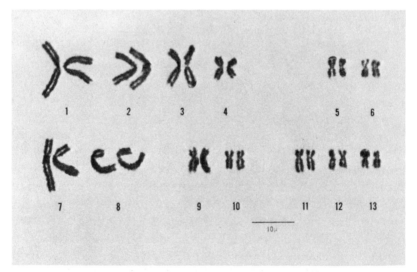

Figure 4-5. Chromosomes derived from a spontaneous renal adenocarcinoma grafted to the anterior eye chamber of a normal adult frog host.
(From DiBerardino, King, and McKinnell, 1963.)

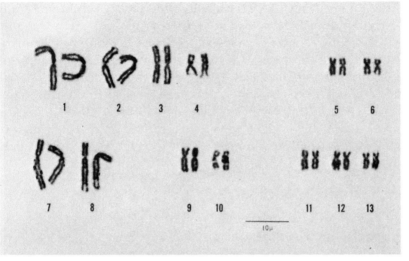

Figure 4-6. Chromosomes derived from an embryo of *Rana pipiens*.
(From DiBerardino, King, and McKinnell, 1963.)

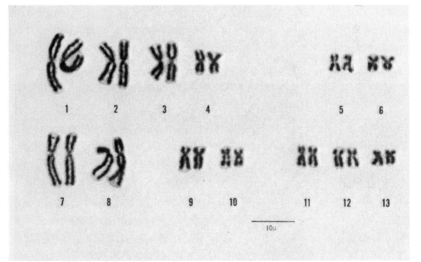

Figure 4-7. Chromosomes derived from normal adult tissue of *Rana pipiens*.
(From DiBerardino, King, and McKinnell, 1963.)

(DiBerardino, King, and McKinnell, 1963). More recently, a study of primary renal carcinomas derived from nine northern Vermont male *R. pipiens* reported that the neoplastic cells are predominately diploid. Karyotype analysis showed that the chromosome constitution of the sampled cells from one tumor was entirely normal. The proportion of normal karyotypes varied in the other tumors from 91% to 25% (DiBerardino and Hoffner, 1969). Thus, it would seem that the challenge of characterizing the renal tumor of the leopard frog by cloning is not precluded by unbalanced chromosome complements that occur so often in cancer.

A Transplantable Tumor

If one desires to insert a living tumor nucleus into an enucleated egg, one must have a reasonably dependable supply of viable tumor. While it is true that spontaneous primary tumors are relatively common among wild-caught *R. pipiens*, it should be noted that leopard frogs are becoming increasingly difficult to obtain (Appendix I). For that reason, it is fortunate that the limited supply of natural tumor material may be amplified by grafting tumor fragments to appropriate hosts. Tadpole tails serve as a fertile field for tumor growth (Figure 4-8). The procedure for grafting is simple

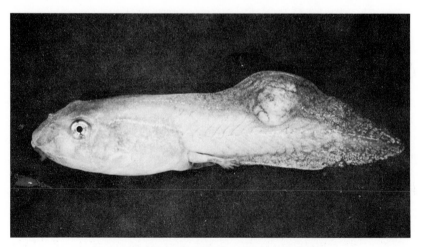

Figure 4-8. Graft of renal adenocarcinoma tissue to
dorsal tail of a *Rana pipiens* tadpole.

(Briggs and Grant, 1943). A small fragment of tumor tissue is inserted into a "tunnel" formed by pushing a glass rod under the epidermis of the tail parallel to the long axis of the embryo. The graft becomes vascularized and in some cases it grows well. Early bilateral thymectomy increases the number of successful tail grafts (Hsu and McKinnell, in press). The anterior eye chamber is perhaps a more useful location for grafting (Figure 4-9). A tumor

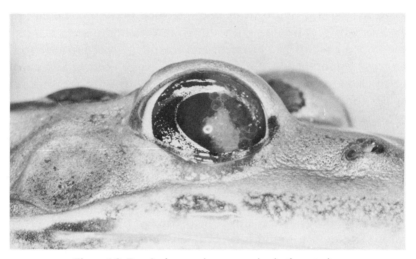

Figure 4-9. Renal adenocarcinoma growing in the anterior
eye chamber of normal adult *Rana pipiens* host.

fragment may be inserted in a slit made at the sclero-corneal junction in an anesthetized host frog (Lucké and Schlumberger, 1940). Growth and survival of tumor grafts to the eye are enhanced with aldosterone treatment of the host frog (Rollins, L., personal communication). The eye location is useful because progress in tumor growth can be monitored visually. Culture in vitro is of course another means of increasing the volume of a spontaneous tumor. Procedures for cell culture of frog tumor-derived material have been developed over a period of years and are described elsewhere (Lucké, 1939; Zambernard, 1973).

The frog tumor has an established viral etiology, a known chromosome constitution, occurs with a high spontaneous frequency, and can be propagated by grafting. The leopard frog was the first vertebrate to be cloned. Surely, if any tumor cell is ripe for characterization by nuclear transplantation, it is the Lucké renal adenocarcinoma. The following records the progress in that difficult arena.

Early Attempts at Tumor Nuclear Transfer

The first report of efforts to reprogram the renal tumor of the leopard frog was that of King and McKinnell (1960). The fact that work is still continuing with the animal tumor system attests to the importance of the problem and the difficulties intrinsic to the system. Spontaneous renal tumors of frogs obtained from the Lake Champlain region of northern Vermont were the source of nuclei for transplantation. Cells of tumors grafted to the anterior eye chamber were also inserted into enucleated eggs. Dissociation of tumor cells was accomplished with either a simple electrolyte solution lacking calcium and magnesium ions or the same solution with an added chelating agent was used. The dissociated tumor cells were drawn into a micropipette and thrust into a previously enucleated host egg. Several cells were drawn into the micropipette because of the relative inexperience of the junior investigator. The novice found it exceedingly difficult to pick up a single tumor cell, and, faced with this plight, was curious to know what would happen if several cells were injected into the recipient cytoplasm. Cleavage, entire or partial, occurred in about one-third of the transplanted eggs (Figure 4-10). Thirteen of the operated eggs became gastrulae and one became an abnormal neurula (Figure 4-11).

Instead of dwelling on the abnormalities of that first tumor nuclear transplant (TNT) embryo, one might wish to consider the point of view of Wilde (1960) who commented, " . . . at least one of these tumor nuclei contained sufficient information to carry on to the point of pretty much setting up almost all of the basic organ structures of a whole embryo, poor though it may be. All that information existed in that nucleus and yet it was a tumor nucleus, and it would seem to me that this is of such fundamental importance that we should all just sit and ponder on it for a minute."

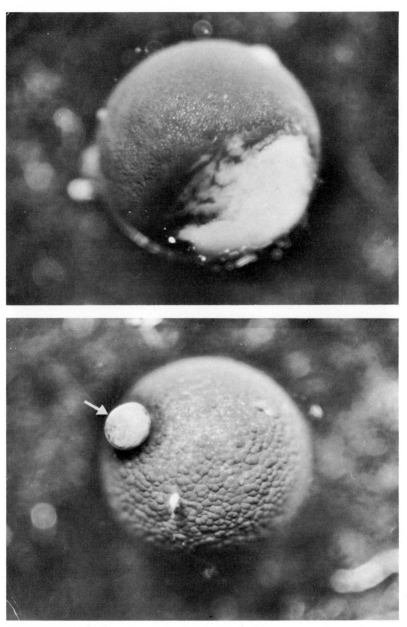

Figure 4-10. (*upper photograph*) Partial blastula formed by the insertion of tumor nuclei into an activated and enucleated egg. The upper two-thirds of the egg is cellular. The vegetal hemisphere material (white) and some darkly pigmented animal hemisphere material in the lower right part of the egg is not cellular. (*lower photograph*) A complete blastula formed by tumor nuclear transplantation. (From King and McKinnell, 1960.)

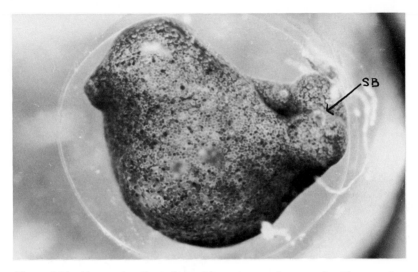

Figure 4-11. Abnormal embryo formed by tumor nuclear transfer. The neurula has not yet hatched from its transparent membrane and suffers from spina bifida (*arrow*). (From King and McKinnell, 1960.)

Later it was shown that at least some TNT embryos are competent to develop to a stage just prior to feeding (Figure 4-12). The best-formed TNT embryos have a head, body, and tail. They swim spontaneously and as a result of prodding with a needle. They are diploid. They have a heart, which may be observed to be beating, and blood corpuscles are seen to be surging through the semi-transparent gills. Histology of these interesting creatures may be compared with normally fertilized tadpoles of the same age (Figure 4-13). It should be noted that TNT tadpole histology appears less normal than illustrations of whole tadpoles. This is due in part to the fact that the TNT embryos were fixed for histology only when it became obvious that they were about to die. A control embryo of approximately 10 days of age (reared at 18°) has well-formed optic cups with spheroidal lenses. The pharynx is flattened dorso-ventrally and the ectoderm is uniformly thin (Figure 4-13 *A*). The tumor nuclear transplant tadpole has optic cups that appear devoid of lenses. The brain, although misshapen, is well differentiated into both gray and white matter. The pharynx is enlarged but surrounded by cartilage as in the control embryo. The ecto-

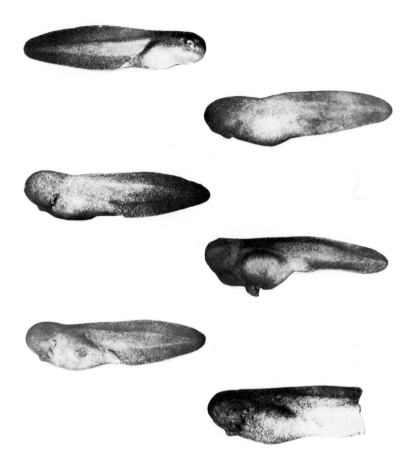

Figure 4-12. Composite photograph showing one normally fertilized control leopard frog tadpole (*upper left*) and five renal tumor nuclear transplant tadpoles. The
embryo of the lower right had its tail clipped for chromosome analysis. It was
able to regenerate its tail. (From McKinnell, 1973b.)

derm is irregular in thickness and has projecting wrinkles in several
areas. Organogenesis is imperfect but histodifferentiation makes
organ identification simple (Figure 4-13 *B*). The hindbrain of a
control tadpole is recognized by its thin roof. In the same area,
the pronephros is well developed with tubules of cuboidal epithelium. Notochord and gut derivatives are easily identified (Figure 4-13 *C*). The same structures are found in a comparable sec-

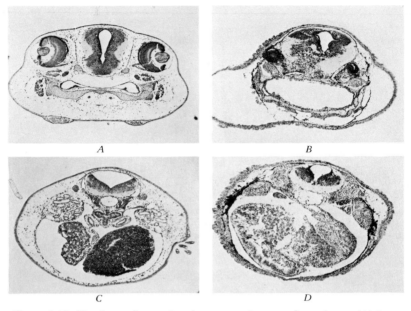

Figure 4-13. Histology of control and tumor nuclear transfer embryos. (*A*) Section through optic cups of normal control embryo. (*B*) Section through optic cups of a tumor nuclear transplant embryo. (*C*) Section through hindbrain and pronephros of normal control embryo. (*D*) Section through hindbrain and pronephros of tumor nuclear transplant embryo. (Photographs (*C*) and (*D*) from McKinnell, 1973b.)

tion of a TNT tadpole but, as in more anterior sections, there are abnormalities. The hindbrain and notochord are minimally deviant. Cells of the pronephric tubules appear to be enlarged and they seem to obliterate the lumena. Ectoderm, as elsewhere, is wrinkled. Yolk-laden cells of the gut occupy most of the body cavity (Figure 4-13 *D*). It should be remembered that despite the described abnormalities, the TNT tadpole has relatively good body morphology and a considerable diversity of tissue types as revealed in histological section (McKinnell, 1962a).

King and DiBerardino (1965) continued these studies by transplanting not only primary and intraocularly grown tumor cells but also nuclei of tumors grown in vitro (Shah, 1962). All of their tumor nuclear transfers developed abnormally with the most advanced experimental animals arresting as abnormal larvae. This study differed from that of King and McKinnell (1960) in that *single* nuclei from tumors were transplanted into each enucleated egg. The results showed that abnormalities of tumor nuclear transplant larvae were not attributable to multiple cleavage centers that might have resulted from the injection of several nuclei. The larvae were mainly diploid but karyotypic analysis revealed chromosomal abnormalities. Arrested post-neurulae, blastulae, and gastrulae were invariably aneuploid (DiBerardino and King, 1965).

The inability of an aneuploid set of chromosomes to interact normally with enucleated egg cytoplasm is of course expected, but what remains unanswered is why tumor nuclei, which are predominately diploid, often become aneuploid when transplanted. Aneuploidy among TNT embryos may result from the same factors that cause many embryonic (i.e., gastrula and older embryo nuclear donors) nuclear transfer embryos to be aneuploid (see discussion in Chapter 3).

Is TNT Development Genuinely Attributable to the
Transplanted Nucleus or Could It Be Due to an
Inadvertently Retained Egg Nucleus?

This question is relevant in any nuclear transplantation experiment in which there is a low yield of cloned embryos. The problem is associated with low yields because of the procedure of manual enucleation of host ova (Appendix II). In the hands of a journeyman embryologist, enucleation is accomplished successfully in excess of 99% of the attempted operations. Indeed King and DiBerardino (1965) report success in 100% of enucleations of inseminated eggs (of 208 operations) and they reported the identification of the egg nucleus in the exovate formed by the enucleation operation. However, the egg chromosomes can occasionally be missed in an attempted enucleation, and it is possible (however unlikely) that the embryonic development that is thought to be attributable to a tumor nucleus may in fact result from gynogenesis. Restitution of the haploid egg nucleus to a homozygous diploid, similar to the diploids described by Subtelny (1958), or fusion of the mature maternal gamete nucleus with the second polar body nucleus would be in both cases consistent with the diploid chromosome number observed in the experimental embryos. The need for suitable nuclear markers in transplantation experiments to eliminate the possibility of genes of maternal origin participating in nuclear transfer experiments has been emphasized by several writers (Harris, 1964; Gurdon, 1968b). I emphatically agree.

Proof of the identity of a transplanted tumor nucleus taking part in development in *R. pipiens* would be possible with the use of a mutant nuclear marker. However, only two (Burnsi and Kandiyohi) have been studied in nuclear transfer experiments (Chapter 2; Appendix I). Regrettably, these mutant genes serve effectively as nuclear markers only in experiments in which the transplant embryos can be reared to metamorphosis. No TNT embryo has survived beyond a pre-feeding tadpole stage (yet). However, grafting procedure and studies of the immune response in frogs will perhaps permit exploitation of these otherwise useful mutant markers (see below).

What other tags or markers are possible if not mutant nuclei? Experiments that provide cytogenetic evidence that a nucleus from a tumor can promote embryonic development will be outlined in the discussion that follows.

Experiments Designed to Provide Cytogenetic Evidence
That Nuclei from Tumors May Be Reprogrammed to
Promote Embryonic Development

The significance of the few TNT embryos that were produced in the early experiments warrants additional experimentation. Techniques are at hand which, if successfully applied to the problem of tumor nuclear transplantation, would provide substantial added evidence that the developmental capacities of tumor nuclei are truly as great as described. It may also be possible to elicit evidence of additional capacity to promote embryonic growth and differentiation that are ordinarily unexpressed in conventional cloning experiments. The operationally feasible experiments involve the use of tumor cells obtained from polyploid and mutant frog tumors combined with well-developed grafting techniques.

Parthenogenetic development may be haploid, diploid, triploid, or a mosaic of several ploidy classes. Parthenogenetic stimulation can result in development directly from the egg nucleus (haploidy) or after karyokinesis with a delay in cytokinesis and fusion of the two daughter nuclei (diploidy) or fusion of the egg nucleus with the second polar body (diploidy). Pure triploid embryos produced by parthenogenesis appear to be relatively rare and are thought to be caused by a delay in cleavage resulting in a diploid chromosome complement that then fused with a retained second polar body (Parmenter, 1933).

A series of triploid TNT embryos developing from transplanted triploid nuclei, with a series of diploid TNT transplant controls that develop as diploids, can only reasonably be interpreted as proliferation of the implanted nucleus without benefit of egg genetic material. For this reason, a source of triploid tumors for nuclear transplantation is of not inconsiderable value.

Herpesvirus-mitochondrial fractions of inclusion-type (i.e., virus containing) tumors have been shown to be oncogenically active in a large percentage of injected prefeeding normal diploid embryos. Methods for preparing oncogenic renal tumor fractions and injection procedures for leopard frog tadpoles are described in Tweedell (1967, 1969) and Tweedell and Wong (1974). It seemed reasonable that if triploid embryos, rather than diploid embryos, were injected with the oncogenic preparation, many of the triploid frogs would develop renal tumors.

Triploid embryos of the leopard frog occur spontaneously (Wright, Huang, and Chuoke, 1976) or they can be produced by a variety of means. These include heat shock (Briggs, 1947), hydrostatic pressure applied to freshly inseminated ova (Dasgupta, 1962), and nuclear transplantation (McKinnell, 1964). Hydrostatic pressure is a convenient procedure for preparing large numbers of triploid embryos in a short period of time.

Triploid embryos for a tumor study were produced by the application of 5,000 pounds/square inch hydrostatic pressure at 4 minutes after insemination. The pressure was maintained for 6 minutes resulting in the repression of the second polar body. The second polar body (1n) and the zygote nucleus (2n) fuse to form triploid (3n) embryos (Dasgupta, 1962). Triploidy in experimental tadpoles was confirmed by aceto-orcein chromosome squashes (DiBerardino, 1962).

A spontaneous bilateral renal adenocarcinoma obtained from a Vermont *R. pipiens* was homogenized and fractionated by differential centrifugation according to the method of Tweedell (1967). Triploid larvae (Stage 25, Appendix VI) were injected with a herpesvirus-mitochrondria fraction after anesthesia. Injections were made through the opercular opening and released intraabdominally. Ten of 23 triploid frogs developed massive renal tumors (Figure 4-14). The histology of the triploid renal tumors (Figure 4-15) did not differ in principle from that of diploid tumors; however, not only was there an increase in chromosomes but nuclear diameter and cell size was larger. Ploidy was ascertained by direct chromosome count (Figure 4-16). The production of triploid renal adenocarcinomas is described at greater length by McKinnell and Tweedell (1969, 1970).

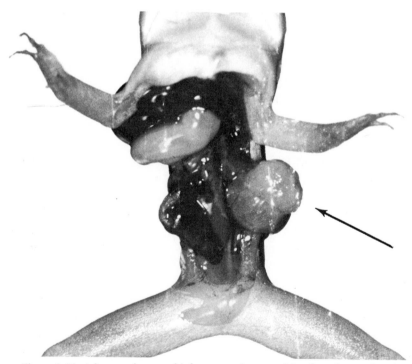

Figure 4-14. Adenocarcinoma of left mesonephros resulting from herpesvirus injection into a triploid tadpole. (From McKinnell, Deggins, and Labat, *Science*, vol. 165, 394-96, 25 July 1969. Copyright 1969 by the American Association for the Advancement of Science.)

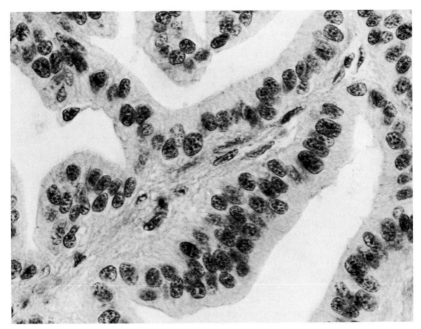

Figure 4-15. Histology of triploid renal adenocarcinoma of the leopard frog,
Rana pipiens. (From McKinnell and Tweedell, 1970.)

Although the study of amphibian polyploid tumors is recent
and continuing, the feasibility of such was suggested over 30 years
ago by Fankhauser (1945). He suggested that after the effects of
carcinogens have been thoroughly studied on diploid cells, the carcinogens should be applied to polyploid cells. In this way, one
could study whether the size of cells or the quantity of chromosome material present has any effect on the process of tumor formation. The production of triploid tumors is a first step in the ful-
fillment of the study suggested by Fankhauser.

Triploid renal adenocarcinomas, produced as described above,
were dissociated and transplanted to activated and enucleated host
eggs. Thirty-three partial and complete blastulae developed from
143 nuclear transfers. Eight blastulae developed to the gastrula
stage, and seven of these hatched and became swimming larvae

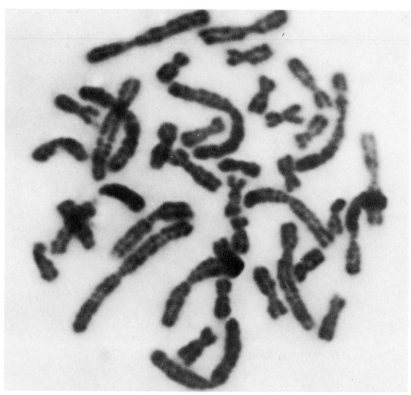

Figure 4-16. Triploid chromosomes (3n = 39) prepared from induced renal adeno-carcinoma by M. A. DiBerardino. (From McKinnell and Tweedell, 1970.)

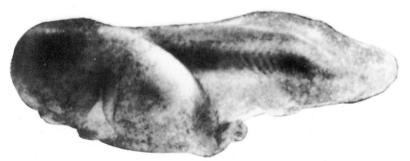

Figure 4-17. Triploid tumor nuclear transplant tadpole produced from transplanting triploid tumor nuclei. (From McKinnell, Deggins, and Labat, *Science*, vol. 165, 394-96, 25 July 1969. Copyright 1969 by the American Association for the Advancement of Science.)

(Figure 4-17). The experimental design is diagrammed in Figure 4-18. The larvae attested to the pluripotency of the triploid tumor nuclei.

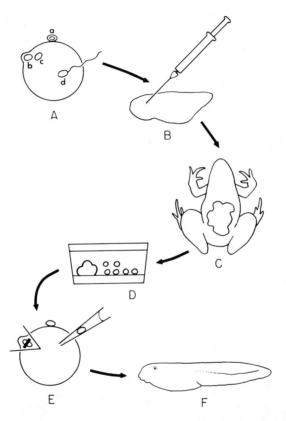

Figure 4-18. Scheme of triploid tumor nuclear transfer. (*A*) Triploidy is induced by hydrostatic repression of the second polar body of inseminated eggs. (*B*) Triploid embryos are then injected with a herpesvirus-containing tumor extract. (*C*) The tadpoles develop into tumor-bearing triploid frogs. (*D*) Triploid tumor cells are prepared for cloning by dissociation in a simple electrolyte solution, inserted into a nucleusless egg (*E*) with a tadpole (*F*) resulting. The triploidy of the tadpole demonstrates the origin of the inserted nucleus, i.e., the tadpole developed as a result of the introduction of a tumor nucleus and not as a consequence of an inadvertently retained maternal nucleus. (From McKinnell and Ellis, 1972a.)

Functional tissue of many types could be observed in the living 3n TNT larvae. As in their diploid TNT counterparts, ciliated epithelium moved the tadpoles in their culture dishes. The tadpoles actively swam when stimulated. Nerve receptors, afferent, central, and efferent neurons, and striated muscle are necessary for swimming in response to stimulation. Cardiac muscle pumped blood cells, which were visible through the thin external gills. "Suckers" in the mouth region secreted abundant mucus. Pronephric ridge, eye primordium, and nasal pits were visible externally as was an open mouth. The body was differentiated into head, trunk, and tail. Tail fin epithelium readily regenerated from a blastema after clipping for chromosome study.

Histological study of the tumor transplant larvae revealed brain, spinal cord, optic cup with lens, auditory vesicle, somites, pronephric tubules, pharynx, midgut, and notochord. An impressive reversion from the malignant state to a condition that at least mimicked the normal was thus obvious in gross and histological examination of the TNT tadpoles.

As stated previously, triploidy is not a flawless cytogenetic marker. Parmenter (1933) reported 3.8% of his parthenogenetically developed tadpoles were triploid. If this percentage is representative of the frequency of triploidy due to gynogenesis, then the probability of the seven triploid TNT tadpoles developing due to parthenogenesis is $P = 0.038^7$ (1.1×10^{-10}). It therefore may be concluded that the embryos developed as a result of the proliferation and differentiation of genetic material derived from a tumor and parthenogenesis is an implausible explanation of our data (McKinnell, Deggins, and Labat, 1969).

With reference to the use of triploidy as a nuclear marker for tumor transplant experiments, it is interesting to note that the procedure was suggested several years earlier. "Un noyau triploide greffé a qualité de noyau marqueur (Gallien, Picheral, and Lacroix, 1963a)."

*Is It Possible to Utilize a Pigment Pattern Mutation to
Identify Mitotic Progeny of a Neoplastic Nucleus?*

In Chapter 2 it was stated that the two leopard frog pigment pattern mutant genes used thus far in nuclear transplantation studies, Kandiyohi (McKinnell, 1964) and Burnsi (Simpson and McKinnell, 1964), are not expressed until metamorphosis. Therefore, the mutant genes would seem to be of little value as genetic tags associated with a donor nucleus since no TNT embryos have yet developed to metamorphosis.

An operationally complex series of experiments may provide a means of exploiting the mutant genes as genetic tags for tumor nuclei, and at the same time, the experiments may provide additional information concerning developmental potentialities not otherwise expressed by the nuclear transfer technique.

Tumor-bearing Burnsi and Kandiyohi frogs would have to be obtained. The tumors from the mutant frogs would then serve as nuclear donors for recipient eggs from wild-type females. The neural crests (containing melanophore-forming cells) from Burnsi or Kandiyohi TNT embryos would then be grafted to normally fertilized wild-type hosts. The expression of the mutant characteristic (mottled or spotless) in the grafts on metamorphosed wild-type hosts would be genetic evidence that the donor nucleus and not the egg nucleus is responsible for the embryonic development. If the experiment were to be successfully accomplished, it would provide information concerning a genetic trait from an adult tumor nucleus expressing itself in grafted *metamorphosed* tissue, a stage in development never yet attained by any part of a tumor nuclear transfer embryo.

In order to perform the proposed experiment, tumors of mutant leopard frogs must be available. Spontaneous Burnsi tumors were reported to occur with a prevalence in excess of 6% in frogs obtained from Minnesota (McKinnell, 1965) and they can be produced by injection of Burnsi or Kandiyohi tadpoles with a tumor fraction containing oncogenic herpesviruses (Tweedell, 1967, 1969).

Rejection of grafts due to histocompatibility factors has been studied in leopard frogs. Neural crest grafts are not rejected if the donor is genetically identical to the host (Volpe and McKinnell, 1966, Roux and Volpe, 1974; see Chapter 7), if an exceedingly large graft is made (Volpe and Gebhardt, 1965), or if the host frog has been rendered immunologically incompetent by removal of the thymus gland (Curtis and Volpe, 1971; Volpe and Turpen, 1976). Since the host in the experiment suggested here would not be genetically identical to the TNT graft, the procedure of Volpe and Gebhardt involving large grafts or the thymectomy procedure of Curtis and Volpe should be utilized to enhance persistence of the grafted tissue.

One crucially important question that would be answered by this experiment is whether or not the grafted tumor transplant neural crest tissue has a genome in a physiological state competent to continue to grow and differentiate on the host frog. There is no way of predicting in advance if the tissue will survive as a graft. The question being addressed here does not involve the immune response of the host (considered above) but rather concerns the viability or potentiality for growth and differentiation of the graft. Parabiosis of euploid endoderm nuclear transfer embryos with normal embryos does not enhance their capacity for growth and differentiation (see Chapter 3 and Briggs, King, and DiBerardino, 1960).

It should be noted, however, that grafting small bits of tissue from inviable (lethal) amphibian embryos to normal host embryos *does* enhance their growth potential. This is true at least in certain tissue types. Thus, brain tissue of hybrid haploid embryos, fated to die at a much earlier stage, undergoes extensive differentiation as an implant in a normal embryo. Haploid epidermis also is not characterized by an autonomous lethality but its primordium develops normally when part of a normal system (Hadorn, 1932). Hybrids between *R. pipiens* and *R. sylvatica* do not develop beyond the gastrula stage, but hybrid ectoderm transplanted to normal *R. palustris* embryos develops neural tissue, optic vesicles, lenses, otic vesicles and pronephric ducts (Moore, 1947). It is thus clear that embryonic tissue fated to an early demise sometimes has substantial latent developmental potentialities that may be provoked into expression by grafting experiments.

Grafts of triploid tissue could be recognized in diploid host embryos because of the larger size of the triploid nuclei and cells and by the increased DNA content. Triploid nuclei contain about half again more DNA than their diploid counterparts (McKinnell and

Bachmann, 1965). Accordingly, triploid tumor transplant embryos could serve as neural crest donors as well as the pigment pattern mutant TNT embryos. Double assurance is provided by a combination of the two markers. A triploid Kandiyohi TNT embryo could serve as a neural crest donor to a wild-type diploid host. The doubly labeled graft would be recognized without question by DNA content and pigment pattern.

Parthenogenesis Induced by Pricking Eggs with
Blood from Triploid Frogs: Its Relevance to
Tumor Nuclear Transplant Embryos

Tumor is composed of epithelial and stromal elements. The connective tissue stroma includes vessels containing blood. It is unlikely that the development of eggs receiving cells from dissociated tumors is due to connective tissue (see next section) but one could argue that the development might possibly be due to the transplantation of a nucleated blood cell (erythrocytes as well as leucocytes are nucleated in amphibia). Supporting this view is the early postulation by Guyer (1907) that parthenogenesis, in at least some cases, is due to the proliferation of leucocyte nuclei in the egg cytoplasm. Is there any reason to believe that any of the tumor nuclear transfer embryos have developed from leucocyte nuclei derived from tumor stroma? In order to answer this question, Dr. Deidre Dumas Labat and I decided to repeat Guyer's experiment using triploid blood, instead of diploid blood, to promote parthenogenesis. Our reasoning was as follows: If parthenogenesis is due in any part to leucocyte proliferation, then a proportion of parthenogenetically developing embryos should have a ploidy corresponding to the ploidy of the blood cells used to promote virgin development. Haploid and diploid parthenogens clearly cannot descend from introduced triploid leucocyte (or erythrocyte) nuclei.

A total of 1,995 eggs were stimulated by pricking with triploid blood cells. Fifty-two blastulae and 43 gastrulae were obtained. Nineteen embryos developed as far as tadpole stages so that chromosome number could be ascertained from tail tip preparations. Fifteen were haploid and four were diploid. None of the parthenogenetically developing embryos were triploid.

We suggest that the parthenogenetic experiments with triploid blood cells provide important evidence that embryonic development is not promoted by introduced blood cells. This is, of course, not an original observation, but in the present context, it is an observation that we believe is of relevance to the developmental potentialities of tumor cells. Thus, in our experiment development was due to replication of the mature gamete nucleus directly (haploid), or it was due to the gamete nucleus after a delay in cleavage with fusion of the first two mitotic descendants of that nucleus (diploid). The authenticity of the tumor nuclear transfers is corroborated by these experiments with gynogenesis (McKinnell, Steven, and Labat, 1976). Splenic lymphocytes of *Xenopus* were recently transplanted (Wabl, Brun, and DuPasquier, 1975). Serial transfer was performed to allow the lymphocyte nucleus to express its full developmental capacity. I presume, therefore, that the first transfer generation of nuclear transfer embryos block at an early stage (if not, why then serial transfer?). If this is true, then while it is possible that an occasional blood cell could be accidentally introduced at the time of tumor nuclear transplantation to the egg cytoplasm, it seems extremely unlikely that the blood cell programmed for the larvae of our tumor cell study.

Are the Nuclei Transplanted from
Tumors Really Tumorous?

Whether or not TNT tadpoles are in fact derived from the genome of a tumor cell is the central question after a decade and a half of experimentation. Klein (1972) stated: "MacKinnell's [*sic*] experiment is obviously extremely important, provided that it can be shown convincingly that the donor cell was a neoplastic cell, rather than a stromal cell." To this same experiment, Gurdon (1974b) echoed: "there is no proof that the cells from which these nuclei were transplanted were tumour cells and not non-malignant host cells present in the tumour tissue."

In fact, the renal tumor of leopard frogs, as all carcinomas, consists of tumor epithelium and a supporting vascularized connective tissue stroma. Hence, there is indeed possibility that host ova may receive transplanted stroma nuclei or blood cells from the stromal vascular network instead of epithelial tumor nuclei. King and Di-

Berardino (1965) believe that the chance of an egg receiving a stroma nucleus is remote because of the low proportion of stroma nuclei to epithelial nuclei. But the chance does exist. Blood cells were discussed in the preceding section.

Stroma, despite its growth sometimes as fast as tumor cells (see McKinnell, 1965, figure 6), is nevertheless not considered to be malignant (Cowdry, 1955). In a sense, the question of stromal versus tumor nuclear potentialities is not crucially important to certain students of developmental biology. Presumably both tissue types are derived from fully differentiated adult tissues. Accordingly, the promotion of embryonic development by either reveals much concerning the functional state of adult DNA and its capacity to interact with egg cytoplasm resulting in coordinated batteries of gene transcriptions.

Stroma seems to be far more resistant to the action of simple dissociation agents (e.g., calcium- and magnesium-free electrolyte solutions with or without a chelating agent, Appendix III) than is epithelium. Certainly the overwhelming bulk of dissociated cells available for transfer are tumor epithelium and not sticky, undisaggregated stromal cells. The observation concerning the difficulty in dissociating stromal cells made in my laboratory leads me to believe, in concurrence with the view of King and DiBerardino, that it is highly unlikely that stroma nuclei are transplanted.

A hypothetical solution to the vexing problem of stromal cell implantation versus tumor nuclear insertion is suggested by a study of salivary tumors in mice induced by the polyoma virus. The question under consideration was whether the tumors had a mesenchyme or epithelium origin. Reciprocally assembled rudiments of the salivary gland were constructed with either epithelium or mesenchyme selectively labeled with T_6 chromosomes. The hybrid salivary rudiments were implanted in hosts that were subsequently challenged with polyoma virus. Tumors that developed were of the karyotype of the epithelial component (Dawe et al., 1971).

The salivary gland experiment suggests a similar approach to the tumor nuclear transplantation problem. It is certainly conceptionally possible (whether or not it is operationally possible is entirely another question) to envision a hybrid mesonephros of diploid stroma and triploid epithelium. Tadpoles with implanted hybrid kidneys could be injected with the Lucké tumor herpesvirus. Tumors that developed after virus injection would serve as nuclear donors for nuclear transfer experiments. The ploidy of the nuclear transplanted tadpole would discriminate between stroma and tumor epithelium as the origin of the inserted nucleus. If stroma is indeed not malignant and if tumor epithelium could be identified unquestionably as the origin of the developmentally significant transplant nuclei, the proposed hypothetical experiment with ploidy-hybrid mesonephros would be of great significance to tumor cell biologists.

Hybridization of radioactive RNA complementary to Lucké tumor herpesvirus DNA to the DNA of Lucké tumor epithelial cells but not to tumor stromal cells would suggest that viral DNA is specific for and limited to the epithelial cells. The feasibility of the method is demonstrated by the detection of a different herpesvirus in human tumor cells by nucleic acid hybridization (zur Hausen and Schulte-Holthausen, 1972; see also Minowada et al., 1974). If viral DNA is indeed limited to Lucké tumor epithelial cells, and if TNT tadpoles contained viral DNA, then the virus would serve as an effective cell marker in cloning experiments. The above comments apply equally to detection of viral gene products (DNA transcripts, Frenkel et al., 1972, or viral specific antigens) as methods for recognizing tumor epithelial cell progeny.

Another method of demonstrating that the tumor nucleus is the progenitor of a tadpole genome would be to culture in vitro colonies of tumor cells from single tumor cells. These clones (the word "clone" is used here to describe a colony of cells descended in culture from a single cell) of tumor epithelial cells should be shown to be tumor by fulfilling as many as possible of the following requirements: tumor cell morphology, loss of contact inhibition, altered nutritional requirements for growth, colony formation in soft agar or methyl cellulose, growth in the anterior eye chamber of an adult host frog, induction of the Lucké tumor herpesvirus by culture at low temperature, detection of virus or tumor-specific antigens, and detection of viral DNA or RNA transcripts (Granoff,

1973). Representatives of the authenticated tumor cell clone could then be microinjected into enucleated host eggs. A ploidy-marked clone (e.g., a triploid tumor cell culture) would enhance these experiments for the same reason that any conventional nuclear transfer experiment needs an identifying tag associated with the donor nucleus.

A direct method of distinguishing tumor epithelial cells from stromal cells is by induced fluorescence with acridine orange (Tweedell, 1965). Fixed primary tumor, fixed tumor cells grown in vitro, and living cultured tumor cells fluoresce with ultraviolet illumination when stained with acridine orange. There is remarkable similarity of fluorescence of cells obtained from the three sources. Tumor epithelial nuclei appear yellow-green and cytoplasm is red to red-brown. Stromal cells have yellow-orange nuclei and a deep green cytoplasm. The acridine orange cytochemical staining procedure is thus diagnostic for epithelium versus stroma. This selective staining procedure was used to identify what kind of cells are dissociated with the gentle methods described above. The majority of cells dissociated from primary tumors and anterior eye chamber grafts fluoresced as epithelial cells (McKinnell, Steven, and Labat, 1976). These studies are continuing.

Another direct method of distinguishing whether or not cultured donor cells are malignant is by their surface morphology as revealed with the scanning electron microscope. Cultured embryonic frog kidney cells are covered with long fingerlike microvilli, which contrast markedly with low ridges and stubby microvilli of Lucké herpesvirus-induced tumor cells (Tweedell and Williams, 1976). An obvious advantage of nuclear transplantation of cultured cells is that the culture can be characterized prior to giving up cells that serve as nuclear donors. The spectrum of cell characteristics currently include acridine orange fluorescence in the ultraviolet, cell surface morphology (both discussed above), and a battery of cytochemical tests (Harrison, Zambernard, and Cowden, 1975).

It should be obvious by this time that several methods are available that should prove fruitful in confirming that tumor nuclei, not stromal nuclei, promote and guide the recipient egg cytoplasm.

Renal Tumor Nuclei are Pleomorphic—Do Both Kinds
of Nuclei Yield Tadpoles When Transplanted?

The pleomorphism of renal tumors (tumors with inclusion-containing nuclei contrasted to those devoid of nuclear inclusions, i.e., virus containing contrasted to "virus-free" tumors) raises the question: Do both nuclear types yield similar results upon transplantation. The question is interesting because it is believed that the inclusion-containing nuclei are fated to die and rarely divide (Rafferty, 1964) and the effect of active virus replication on host cell developmental potentialities is thus an intriguing question. Strangely enough, it is also unanswered. It would seem probable that inclusion nuclei can promote little if any development of a host egg; however, studies thus far of diploid tumor nuclear transplantation unfortunately have failed to report which nuclear type was transplanted. Tumors induced in triploid frogs (discussed above) are known to be of the non-inclusion type renal adenocarcinoma. Thus, the cloning experiments with cytogenetically marked tumor nuclei are of a known nuclear morphology.

Epigenesis and Frog Renal Tumors

The tumor nuclear transplantation experiments reviewed above suggest that the frog renal tumor has arisen by some variant process of differentiation that has left the tumor cell genome unaltered from that of a zygote nucleus,—unaltered, that is, with respect to its capacity to direct recipient nucleusless cytoplasm to form a swimming tadpole. However, somatic mutation is not excluded because, although it now seems well established that the tumor genome is pluripotent, it has not been demonstrated that it is totipotent. Mutation to one or more loci could inhibit the development of a vital organ to the fully differentiated adult state. This possibility has not been ruled out by the present experiments.

Pluripotency of the frog renal tumor was suggested by the regeneration experiments of Rose (1949) in salamanders several years before the nuclear transfer experiments were started. Frog tumor tissue was implanted to the limb of a salamander. The limb was amputated after the tissue began to grow. The urodele host to the frog tumor tissue regenerated a new limb. Among the muscle and cartilage cells of the host's new limb were nuclei thought to be frog. The conclusion that a tumor cell nucleus can be reprogrammed to direct the formation of a tadpole is in harmony with the inference of tumor cell pluripotentiality derived from the earlier regeneration experiments.

Episomes are exogenous genetic material that becomes incorporated in the genome of a host cell. The foreign DNA may replicate in an integrated state with the host DNA, or it may become more or less independent of the host cell and the episome DNA may increase while the DNA of the cell remains dormant or disintegrates. Episomes and some systems in which they are thought to function are described at length in a book by Campbell (1969). What effect does latent herpesvirus DNA have on expression of frog genetic material when studied by the cloning procedure? There will be no answer to that question until someday a frog develops from a tumor cloning experiment, if that ever happens. A frog would attest to the inability of the viral DNA to inhibit activation and expression of the entire frog genome. If we are not able to obtain a frog by tumor nuclear transfer, then for the time being at least, we will be unable to distinguish between the alternatives of imperfect development due to differentiation of kidney cells and imperfect development due to viral DNA interference with the expression and activation of frog kidney cell DNA.

I hope that this chapter has made it perfectly clear that we have no thoughts of closing the door on further experimentation.

CHAPTER 5

The Transplantation of Nuclei Obtained from Embryos beyond the Blastula Stage: *Xenopus laevis,* Urodeles, and Some Studies from Russia and China

The nuclear transplantation technique devised for the northern leopard frog, *Rana pipiens*, by Briggs and King (1952) was adapted for other anuran species within a few years. That the cloning procedure was readily applicable to forms other than *R. pipiens* was shown by the variety of anuran and urodele species that were studied following the first amphibian cloning report (see Appendix IV for a list of amphibian species used in nuclear transplantation studies). While it is true that many amphibian species have been used in cloning, the analysis of developmental changes in nuclear totipotency, as assessed by nuclear transplantation, has been studied extensively thus far primarily in *Rana pipiens*, the leopard frog, and in *Xenopus laevis*, the South African clawed frog (Figure A-32). The results obtained utilizing embryonic nuclei of the latter species will be considered in this chapter together with some urodele nuclear transfer studies and cloning studies from eastern Europe and the Orient.

Procedural Differences in the Utilization of Rana *and* Xenopus

In order to evaluate the significance of differences that may result in cloning experiments utilizing different species, methods used in obtaining the results should be compared. Thus, if marked experimental differences are reported, one must seek to distinguish if the dissimilarities in results are conditional and due to technical factors or if they are intrinsic and result from biologic factors.

The nuclear transplantation methods in *Rana* and *Xenopus* differ primarily in the preparation of the recipient egg (discussed in greater detail in Appendix II of this volume and in Gurdon, 1974b). The *Xenopus* egg was neither activated nor enucleated in early experiments (Fischberg, Gurdon, and Elsdale, 1958a, b), but an enucleation procedure was subsequently adopted. Manual extraction in *Xenopus* of the second meiotic division spindle and chromosomes with a glass microneedle is possible technically but is hampered by poor viability of operated eggs (only about 27% survive to hatching, Hamilton, 1957). Ultraviolet inactivation of maternal nuclear material in *Xenopus* ova has been shown not only to be reliable but the irradiation effects a change in the egg surface in such a way that a micropipette can be inserted into the egg with minimal trauma (Gurdon, 1960a). The recipient egg is dry when it receives the donor nucleus, i.e., a solution does not cover the eggs at the time of nuclear transplantation. A nucleolar marker associated with the donor nucleus provides cytogenetic evidence that the transplanted nucleus is responsible for the development of the recipient egg (Elsdale, Fischberg, and Smith, 1958). The cloning procedure developed and used by Gurdon is diagrammatically illustrated (Figure 5-1). It would seem that although egg and donor nuclei are manipulated in a somewhat different manner from that of the Briggs and King procedure, the methods are in fact quite parallel, and accordingly, the results obtained with *Rana* and *Xenopus* ought to be comparable.

How Similar Are the Results Obtained by
Nuclear Transfer in Xenopus *and* Rana?

Nuclear transplantation studies in *Xenopus* are indeed similar to those reported for *Rana*. This fact was graphically illustrated by Gurdon (1963). There are fewer nuclei from more advanced stages that are able to promote complete blastula formation in both *Xenopus* and *Rana* (see Gurdon, 1963, Figure 2). The proportion

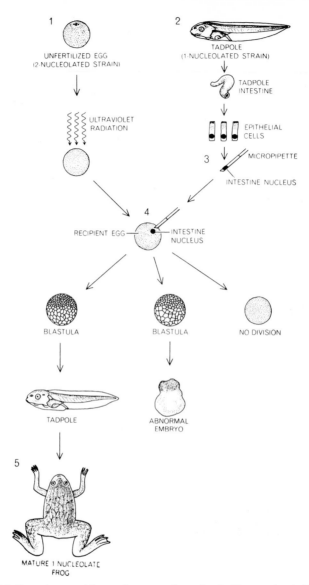

Figure 5-1. Procedure used for nuclear transplantation in *Xenopus laevis*. The wild-type (2-nucleolar type) recipient egg (*1*) is irradiated with ultraviolet light, which effectively removes the maternal chromosomes. A cytogenetically distinct (1-nucleolar type) donor embryo (*2*) provides cells whose nuclei are drawn into a micropipette (*3*). The nucleus-containing micropipette is then thrust into the previously irradiated ovum (*4*). In some instances, no division results; in other cases, an abnormal embryo ensues, and in still other operations, a nuclear transplant frog grows from the operated egg (*5*). (From Transplanted nuclei and cell differentiation by J. B. Gurdon. Copyright © 1968 by Scientific American, Inc. All rights reserved.)

of transplanted nuclei promoting normal cleavage in the leopard frog drops precipitously with development of the donor embryo beyond gastrulation in contrast to a less drastic reduction in proportion of nuclei of older stages competent to promote cleavage in *Xenopus*, but the yield *drops with increased age of donors in both species*. Similarly, progressively fewer blastulae obtained from transplanted nuclei of later embryonic stages are competent to continue development as normal tadpoles in both *Rana* and *Xenopus* (Figure 5-2). The descending curve showing reduced developmental capacity of older nuclei from *Rana pipiens* is steeper than the curve for *Xenopus laevis*. Thus, it would seem that the African

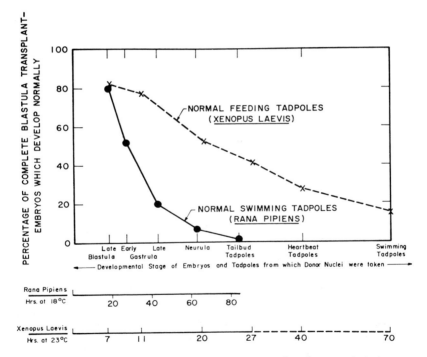

Figure 5-2. Fewer blastulae, produced by nuclear transfer of progressively later stage donors, are competent to develop normally in both *Xenopus* and *Rana*, as shown in the graph. A significant number of tadpoles develop from blastulae after the transfer of donor nuclei of 70-hour-old swimming tadpoles in *Xenopus* in marked contrast to *Rana*. (From Gurdon, 1963.) Compare this graph with Figure 5-3 showing the same data plotted with the age of donor nuclei expressed in hours rather than stages.

clawed frog is different from the North American leopard frog when characterized by nuclear transfer. However, the reader is urged to note the legend to the graph relating to hours of development of both *R. pipiens* and *X. laevis* at the abscissa. Observe that in this graph, *R. pipiens* tailbud embryos are already 80 hours old while swimming tadpoles of *X. laevis* are but 70 hours old. I replotted this data using *age* of the nuclear donor in place of *stage* of the nuclear donor. The curves become similar when compared in this manner (Figure 5-3) (McKinnell, 1972). The suggested manipulation of the abscissa of cloning data on two species of frogs probably is not dissimilar to the deformation of Cartesian coordinates designed to reveal similarity in anatomy of different forms by Thompson (see Thompson, 1942, chapter XVII, "On the theory of transformations, or the comparison of related forms").

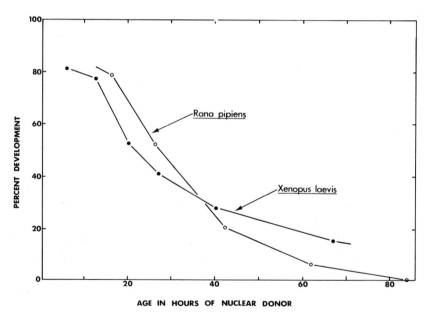

AGE IN HOURS OF NUCLEAR DONOR

Figure 5-3. Graph illustrating the decline in the percentage of nuclear transplant blastulae which develop normally associated with increasing age of nuclear donor. The graph is prepared from the same data as that illustrated in Figure 5-2 but the abscissa has been altered to chronological age instead of embryological stage thereby revealing striking similarities in the capacity for development of *Xenopus* and *Rana* nuclear transplant embryos. (From McKinnell, 1972.)

While one may be tempted to manipulate data to reveal underlying similarities, one can be misled easily in his ambition to seek
common modes of development. Thus, one may question if it is
better to compare stage or age of nuclear donor. Perhaps a more
fundamental question relates to how similar *Xenopus* and *Rana*
are with respect to considerations other than nuclear transfer. A
comparison of their biology is perhaps in order. The comparison is
motivated by a desire to obtain a perspective for comparing cloning data. In a very real sense I am seeking an answer to the question of the biological significance of declining potency associated
with increased age of cloned nuclei.

R. pipiens *and* X. laevis *Compared*

It is interesting that *R. pipiens* and *X. laevis* should have been
the subject of intensive nuclear transfer research because the two
anuran species differ greatly (Rose, 1950; Goin and Goin, 1962).
The leopard frog can survive arid conditions. It is terrestrial during
the summer season in northern United States. I have collected mature leopard frogs on dusty roads and fields a half mile or so from
the nearest standing water in Minnesota and Vermont. Xeric seems
an appropriate adjective to describe the habitat of an anuran that
creates a small cloud of dust as it jumps along a dry path on a summer afternoon.

The South African clawed toad, in contrast, is fully aquatic before and after metamorphosis. It makes its home in lakes, ponds,
swamps, and wells over much of southern Africa and remains sub-
merged most of the time. *Xenopus* has an elaborate system of lateral line organs associated with its aquatic existence. The platanna
(as the South African clawed toad is known in its home range)
buries itself in mud when ponds become dry.

There does not seem to be agreement concerning the phylogeny
of frogs. Accordingly, it is imprudent to unequivocally state that
one group is primitive and another group more specialized. However, it is safe to state that in most schemes of anuran evolution,
the Pipidae (the family of *X. laevis*) is considered to be more primitive (archaic) than the Ranidae (the family of *R. pipiens*) (Inger,
1967, Lynch, 1973).

What are some of the characteristics that have led biologists to consider the Pipidae to be more primitive than the Ranidae? The characteristics are of several kinds, involving a consideration of the fossil record, larvae, adult morphology, and physiology.

In contrast to what is popularly believed, there is a fossil record of frogs. Nearly 900 fossil pipids were found in Israel that date from the early cretaceous (about 135 million years ago). The fossil pipids do not differ strikingly from extant pipids and the fossils are believed to have been aquatic as is *Xenopus* now (Nevo, 1968). Thus, the Pipidae clearly have an ancient origin and, further, they seem to be related to the only known fossil frog family, the Palaeobatrachidae (Spinar, 1972). Thus, *Xenopus*, the most primitive (Estes, 1972) of the living and primitive pipids, is an anuran that is classified in a family that is ancient and exhibits structures indicating a relationship with an extinct group.

Fossil pipids, like living pipids, had ribs as larvae. The presence of ribs in anurans is considered to be primitive. Ranids are bereft of ribs at all stages of development.

Developmental divergence between *Xenopus* and *Rana* has its onset at mating. Amplexus in frogs is a sexual embrace or clasping that permits sperm to be delivered in close proximity to ova as they are being spawned. Male *Xenopus* have a pelvic clasp (Figure 5-4) considered by some to be primitive in contrast to a pectoral clasp during amplexus of *Rana* (Figure 5-5) and other more advanced frogs. Development in the South African clawed toad is substantially faster than in the North American leopard frog. It takes twice the time for the American frog to reach the early gastrula stage (Stage 10, Appendix VI) as it does for the South African clawed toad to reach the same embryonic stage (Stage 10¼, Nieuwkoop and Faber, 1967). Hatching from jelly membranes takes place with dispatch in *X. laevis* in 24 to 48 hours (Elkan, 1957) in contrast to the 75 hours for the more dilatory *R. pipiens*. The differences in developmental rate cited here are substantially less than would be observed if *R. pipiens* was reared at its customary laboratory temperature of 18°C. The data for the leopard frog have been calculated for 23°C (the customary laboratory temperature for *Xenopus*) to make the comparison more valid. An elevation of temperature increases developmental rate in *R. pipiens*

Figure 5-4. Primitive style of amplexus characteristic of *Xenopus*. The male is smaller than the female and grasps her in the pelvic region immediately above the posterior limbs. (From *Xenopus: The South African Clawed Frog* by E. M. Deuchar. Copyright © 1975, John Wiley & Sons. Reprinted by permission of John Wiley & Sons, Inc.) Compare this primitive mode of sexual embrace with the more advanced pectoral clasp of *Rana* illustrated in Figure 5-5.

Figure 5-5. The ranid male clasps the female in the pectoral region under her anterior limbs during amplexus. (From Aronson, 1943.) Compare this mode of sexual embrace with that of *Xenopus* (Figure 5-4).

(Moore, 1939), but despite the more rapid growth of the leopard frog at the high platanna temperature, it still is much slower than *X. laevis*.

Tadpoles of anura are of four types: type 1 has soft mouth parts and two spiracles; type 2 has soft mouth parts but has a single median spiracle; type 3 has a single median spiracle but has horny mouth parts; and type 4 has horny mouth parts and a sinistral spiracle (Figure 5-6). Type 1 is considered primitive and it is characteristic of *X. laevis*. Type 4 is specialized and is characteristic of *R. pipiens* (Orton, 1953).

Musculature and details of vertebral development serve to illuminate the fact that *Xenopus* and its relatives are different from *Rana* and its kin. These and other characteristics that may be useful in studying the evolution of anurans are discussed in Griffiths (1963) and Inger (1967).

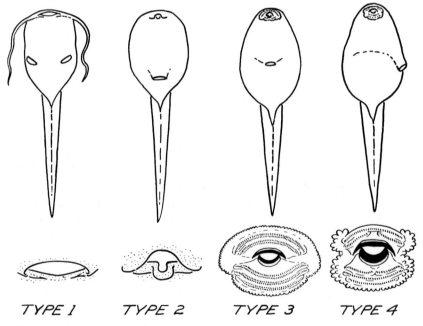

TYPE 1   TYPE 2   TYPE 3   TYPE 4

Figure 5-6. Diagram of tadpole morphology. *Xenopus* has a Type 1 tadpole which is considered primitive. The tadpole of *Rana* is Type 4 which is specialized. (From Orton, 1953.)

Regeneration experiments, more than any technique other than nuclear transplantation, manifest the genetic and developmental potentialities of somatic cells. Here, as in anatomy and development, are found substantial dissimilarities. Regeneration of limbs in adult leopard frogs can be provoked only with the most heroic efforts, with the resulting regenerate only poorly resembling the normal limb (Rose, 1970). The capacity to regenerate limbs in *Xenopus* is not completely lost even in the adult (Beetschen, 1952; Goode, 1967; Dent, 1962) and there is considerable regenerative ability in the central nervous system of *Xenopus* (Srebro, 1959). The cerebellum of *Xenopus* regenerates in the larva and after metamorphosis (Filoni and Margotta, 1971). One means of distinguishing one South African clawed toad from another is by toe-clipping but the claws regenerate in a short time and must be reclipped at least every 2 months in growing immature toads (Gurdon, 1967b). It has been said that *Xenopus* seems more like the salamander than it does other anurans in its capacity to regenerate missing organs (Hay, 1966).

Cold-blooded vertebrates reject grafted tissue from other individuals of the same species as do higher vertebrates (Chapter 7). Both *X. laevis* and *R. pipiens* reject allografts but the leopard frog responds to allografts in an acute manner similar to mammals while *X. laevis* resembles urodeles in its chronic reaction to an allograft (Cohen, 1971).

The male of *R. pipiens* is the heterogametic sex while the reverse is true in *Xenopus laevis* (Nace et al., 1974). *Xenopus* lacks a tongue and movable eyelids. It has no distinct tympanic membrane. *R. pipiens* has tongue, eyelids, and tympanum.

The purpose of the discussion above is not to become engaged in the polemics concerning the relative positions of *X. laevis* and *R. pipiens* in an anuran phylogeny. However, it is my purpose to emphasize that in many ways the two species are exceedingly different. In this sense, I do not concur with Inger (1967) who paraphrased Gertrude Stein: "a frog is a frog is a frog."

Now, if cloning data between the two species were found to be disparate, I would suggest that the inconsistency in data probably is related to the rather substantial species distinctions described above and is not of fundamental biologic importance. However, because nuclear transfer results are so similar in such widely distinct anuran species as *X. laevis* and *R. pipiens*, it indicates to me that the data from the experiments may be of a general nature and perhaps reflect processes held in common by many or all higher organisms.

Embryological Comparison Expressed in
Terms of Age Instead of Stage

It is certainly common practice among embryologists to compare embryological stages instead of embryological ages. The practice has merit in chick development because newly laid eggs may have had variable periods of incubation in the hen. Chick embryos of similar anatomical development are probably more comparable in their physiology than are chick embryos of different embryological stage but of equivalent incubation outside of the hen. The comparison of stage rather than age may lose validity if certain biochemical processes are correlated. Thus, the biochemistry of protein synthesis is dependent on substances that have a finite existence. Indeed, the half-life of a number of ribonucleic acids has been calculated and some are known to be short. For example, unstable nuclear RNA has a half-life measured in minutes (Soeiro et al., 1968; Brandhorst and Humphreys, 1971). It seems reasonable to relate differentiation to biochemical substances that are important to differentiation instead of to arbitrary morphological stage. These substances have finite lives.

To elucidate, consider the effect of actinomycin on early development of different species. Actinomycin, which inhibits transcription of DNA, causes sea urchin embryos (Gross and Cousineau, 1964; Summers, 1970) and amphibian embryos (Brachet, Denis, and de Vitry, 1964; Wallace and Elsdale, 1963) to block at late blastula or early gastrula stage. There is an interesting exception to the generalization that actinomycin blocks embryonic development at the late blastula or early gastrula stage. The exception is the case of ascidian eggs treated with actinomycin. Ascidian embryos will develop to swimming larvae in the presence of actinomycin (Reverberi, 1971). How can the development of ascidian embryos be related to development of sea urchins and frogs? The answer to the question may lie in the fact that it takes the same time for ascidians to reach larva stage as it does for sea urchins and frogs to become blastulae. Thus, actinomycin is probably no less effective nor any slower in blocking transcription in ascidians than it is in frogs, but the apparent difference in results may relate to chronological considerations such as the immensely greater speed of development of the former and the tardiness of development of the latter.

Another related series of observations concerns the onset of ribosomal RNA synthesis in embryos. Once again, sea urchins and frogs are similar in that both start ribosomal RNA synthesis at the onset of gastrulation (Davidson, 1976). Mammals seem to be precocious in this respect for early cleavage stages may be involved in ribosomal synthesis. Incorporation of RNA precursors into nucleoli starts just after the four-cell stage in the mouse (Mintz, 1964). It should be noted here that by amphibian standards cleavage is slow in mammals. Thus, the two-cell stage in the rabbit occurs 22 to 24 hours after mating or 10 to 12 hours after fertilization and the four-cell stage occurs 13 to 18 hours after fertilization (Nelsen, 1953) and the four-cell stage in the mouse may not occur for up to 50 hours after insemination (Rugh, 1968). If one considers that mammals develop somewhere near 37°C while amphibia and sea urchins develop at a temperature 10 to 20 degrees cooler, then the "precociousness" of ribosomal RNA synthesis in mice seems less impressive. When ribosomal RNA synthesis is considered from the age of its onset instead of the stage of its onset, mammals do not appear to be so dissimilar to amphibia.

For these reasons, I believe that an age comparison of cloning experimentation may be more revealing than a stage comparison of the very different *Xenopus* and *Rana*. In summary, it would seem that the South African clawed frog and the northern leopard frog are biologically very different; yet, when cloning of the two species is compared by age of nuclear donor, the data are essentially identical. It is therefore concluded that age dependent changes in nuclei, as revealed by nuclear transplantation, are probably real and are of the same nature in both species. A detailed account of nuclear transfer experiments in *Xenopus* follows.

Differentiating Cells: Endoderm

It would seem that Briggs and King and their associates became committed early to a careful study of the first changes in nuclei that can be detected by the cloning technique. Accordingly, during the decade and a half after 1952, these investigators were characterizing by a variety of studies the nature of early restrictions in developmental potentialities that were observed in some transferred nuclei (Chapter 3). On the other hand, Gurdon seemed to have been concerned with the exceptional nucleus from older embryos (i.e., he sought to find nuclei, from more advanced embryonic stages, in which the genome appears to remain totipotent).

The developmental capacity of transplanted nuclei obtained from columnar epithelial gut cells of feeding tadpoles (stages 45-48, Nieuwkoop and Faber, 1967) is reported in Gurdon (1962a). The gut cells have a striated border and are stated to be "fully differentiated" (Gurdon, 1962a; Gurdon and Uehlinger, 1966; Gurdon and Graham, 1967); yet they can be recognized by "their possession of yolk" (Gurdon and Uehlinger, 1966), the quantity of which is variously described as "a few yolk platelets" (Gurdon, 1962a) or "large content of yolk" (Gurdon, 1968b). The gut cells "are the only larval intestine cells which have not resorbed all their yolk" (Gurdon, 1974b). The nuclear donor tadpoles had just begun to feed (stages 46-48, Nieuwkoop and Faber, 1967) and probably varied in age from about 4½ days to 7½ days if reared at 22° to 24°C. The hind limb anlagen first appears at stage 46; the forelimb rudiment appears as a bud at stage 48. The tadpole tail does not resorb until about 58 days after fertilization and sexual maturity requires about 1 to 2 years of growth. The descriptions concerning yolk content of the "fully differentiated" donor cell and the age in days of the nuclear donor are included lest some reader equate "fully differentiated" with adulthood, which takes about a month to a year or so depending upon what is meant by an adult.

Ten normal tadpoles were obtained from 727 nuclear transplantations. The 10 larvae comprise a major accomplishment. Yet, Gurdon does not believe the yield (1½%) is indicative of the developmental potentialities of the majority of the nuclei that were transplanted. He discounted the significance of the low yield by

an examination of his transplant "failures." The 10 normal tadpoles ensued from first generation transplants. Gurdon found that
some nuclei that promote abnormal development after first transfer promote normal development with serial transfer. Thus, the
nuclei from some partial blastulae resulting from tadpole intestine
nuclear transfer could when retransplanted develop to the normal
feeding tadpole stage. "The combined results from first transfers
and of serial transfers demonstrated that at least 7 percent of the
intestine nuclei possessed the genetic information required for the
formation of normal feeding tadpoles" (Gurdon, 1962a).

Forty-eight percent (347 of 726 attempts) showed no cleavage
at all. Cytological examination of these failures indicated that the
transplanted nucleus was not effectively exposed to the host cytoplasm or was at an inappropriate mitotic stage; Gurdon argued
that these cases should be excluded from the results. If they are
excluded, the gut nuclei competent to promote formation of feeding tadpoles then constituted 24% of the remaining successful
transfers. A similar interpretation of the data suggests that 70% of
transplanted nuclei are competent to program for development to
a muscle response stage (an earlier stage than feeding tadpoles)
(see Gurdon, 1962a, table 5).

A totipotent nucleus must not only promote the formation of
a tadpole when transplanted but must be competent to give rise to
cells that develop into a sexually mature animal. Nuclear transplant frogs were mated to frogs not derived from transplantation
to ascertain the ability of an implanted nucleus to promote functional germ cell formation. Sexually normal transplant frogs were
reported. This showed that some cloned nuclei, obtained from
blastulae and later stage donors, have not undergone changes that
prevent the normal expression of that portion of the genome necessary for gametogenesis (Gurdon, 1962b). "Fertile" intestine nuclei
have also been reported. Sexually mature frogs, competent to produce mature gametes, were produced by transfer of nuclei obtained from feeding tadpole-gut cells (Gurdon and Uehlinger,
1966).

Genuinely adult nuclei have been transplanted by Laskey and
Gurdon (1970) and Gurdon, Laskey, and Reeves (1975). Those

studies have not, as of this writing, resulted in normal development of the recipient egg. The limited development of adult *Xenopus* nuclear transplant embryos is discussed in Chapter 6.

Differentiating Cells: Neural Tissues

There is a substantial reduction in the percentage of late gastrulae neural nuclei that sustain development as far as feeding larvae after transfer. Nevertheless, normal larvae developed from transplanted presumptive brain nuclei derived from the closed neural fold stage in 5 of 221 transfers (Simnett, 1964b). In contrast to the specificity of deficiencies reported in a few cases in leopard frog cloning of neural nuclei (Chapter 3), the abnormalities in deficient embryos that developed from presumptive brain nuclei are identical with those of abnormal endoderm nuclear transfer embryos (Simnett, 1964b).

More recently it was reported that neural nuclei from the forebrain of advanced neurulae (Stage 22, Nieuwkoop and Faber, 1967), although not competent to promote development beyond the blastula stage directly (first transfer generation), will sustain development to the adult frog in subsequent (serial) nuclear transplantations (Brun and Kobel, 1972).

Differentiating Cells: Other Tissue Types

Epidermal nuclei from hatching tadpoles (Stages 36-42, Nieuwkoop and Faber, 1967) when inserted into an enucleated egg will produce an occasional embryo that develops as far as gastrulation (2 of 440 attempted nuclear transfers). However, blastulae produced by the epidermal nuclear transfers may serve as nuclear donors, which results in the recipient egg developing to an adult frog. Thus, while epidermis results in no frogs at the first transfer generation, serial nuclear transfer enhances the capacity of the nuclei to program for normal development. Nineteen frogs were obtained from 577 serial nuclear transplantations (Brun and Kobel, 1972). The donor nuclei were marked appropriately with the nucleolar mutation (Elsdale, Fischberg, and Smith, 1958).

Skin melanophores of advanced larvae (Stage 56, Nieuwkoop and Faber, 1967), in first nuclear transfer generation or by serial nuclear transfer, never promote development beyond an exceptional gastrula. However, nuclear transfer of in vitro cultured melanophores from younger larvae (Stages 35-40, Nieuwkoop and Faber, 1967) produce tadpoles (Kobel, Brun, and Fischberg, 1973). Tadpoles are also produced by nuclear transplantation of aneuploid cell line A-8 (Rafferty, 1969) derived from the liver of an adult *Xenopus* (Kobel, Brun, and Fischberg, 1973). Nuclear transfer of cultured adult cells is considered below and again in Chapter 6.

Blastema Nuclear Transfers

Regeneration studies of amphibian limbs were surveyed briefly in this chapter. Certainly, *Xenopus* as an adult regenerates many tissues well. Because of the regeneration studies, it would be extremely interesting to know the capacity of a nucleus from a blastema cell to promote differentiation.

There are few published records of blastemal cells implanted to enucleated eggs; they include Burgess (1967, 1974) with *Xenopus* and Dasgupta (1969, 1970) utilizing urodeles. Burgess amputated the hind limb midway on the tibio-fibula of larvae (Stages 59-60, Nieuwkoop and Faber, 1967). The operated limbs were allowed to regenerate for one week. She then transplanted blastema cell nuclei both singly and in clumps of several cells. The results are disappointing because blastema cell nuclei when transplanted singly gave only irregular or abortive cleavages. However, a few blastulae formed and 2 early gastrulae developed (from 186 transfers) as the result of implanting clumps of blastema cells. The author suggests that the superior results obtained by the clump method over single transfers may be due to the lack of nuclear protection because of too little cytoplasm in the single isolated cells. Whatever the cause of the scanty development, it may be concluded that either the nuclear transplantation method does not comprise an adequate test system for blastema nuclei (we know that more genetic information is resident in their nuclei than is revealed by cloning; see review by Rose, 1970) or the procedure was not optimally applied to the blastema cells. Gurdon, when describing the nuclear transplantation procedure in *Xenopus*, has stated that thorough disaggregation of donor cells is essential for normal development (Gurdon, 1960b). Burgess got her best results with clumps of cells. Additional experiments are needed to clarify to what extent blastema nuclei are competent to promote development of nucleusless eggs of the South African clawed frog. Urodele blastema nuclear transfers (Dasgupta, 1969; 1970), are discussed later in this chapter.

The Transplantation of Nuclei from Cultured Cells

Nuclei from monolayer cell cultures of embryonic and adult cells have been cloned. A cell culture originated from cells of swimming tadpoles (Stage 40, Nieuwkoop and Faber, 1967) provided nuclei, which when transplanted to enucleated eggs gave rise to one first transfer and two serial transfer embryos that grew to become frogs (Gurdon and Laskey, 1970a). The paper is of greater significance with respect to procedure than it is with respect to results. It should be noted that the embryos that provided cells for culture (Stage 40, Nieuwkoop and Faber, 1967) were younger than the embryos that provided the yolk-laden gut cells whose nuclei were transplanted with no prior culture (Stages 46-48, Nieuwkoop and Faber, 1967) and some of which were shown to be developmentally totipotent (Gurdon, 1962a). Thus, the capacity to program for total embryonic development informs us that the culture procedure did not diminish the potencies of nuclei derived from a stage younger than one already known to retain at least some developmentally totipotent nuclei. The methodology for transplantation of nuclei of cells cultured in vitro is described in Gurdon and Laskey (1970b).

Once a technique has been developed such as the transplantation of cultured cell nuclei, it can be further exploited to answer additional questions. One such question concerns whether or not the genome of cultured *adult Xenopus* cells are in the physiological state such that they have the capacity to order egg cytoplasm to form other normal sexually mature frogs. This is the question that was asked by some of the early manipulators of embryos (Chapter 1). It is an exceedingly important question and will be considered again at length in Chapter 6.

Modulation of Transcriptional Repertoire
by the Cytoplasmic Environment

It was noted in Chapter 3 (see especially Figure 3-3) that different stages of development are characterized by specific patterns of nucleic acid synthesis. For example, we have known for some time that frog blastulae are not noted for ribosomal RNA synthesis but with the onset of gastrulation there is rapid synthesis of new rRNA (Brown, 1967). One may wish, therefore, to ask if a nucleus that already has manifested a nascent biochemical specialization (i.e., a nucleus that is vigorously synthesizing ribosomal and all other major classes of RNA) will continue to behave in its characteristic manner when placed back into the less specialized environment of egg cytoplasm. In other words, will the egg cytoplasm repress genomic activity that has been previously activated? One may also ask if the manipulation can be managed in a physiological manner so that both the environment (cytoplasm) and the reacting genome remain viable.

The answer to these questions is yes. No RNA synthesis can be detected within 20 minutes of transplantation of a neurula nucleus. Further, the mitotic descendants of the nucleus that was previously transcriptionally active remain quiescent with regard to RNA synthesis through much of cleavage. Between mid- and late-cleavage, heterogenous nuclear RNA synthesis begins followed by 4s RNA (a low molecular weight RNA), and ribosomal RNA synthesis begins at the onset of gastrulation. The patterns of RNA synthesis in the neurula nuclear transplant animal are identical to those of fertilized controls (Gurdon and Brown, 1965; Gurdon and Woodland, 1969). Thus, specific nucleotide sequences can be reversibly repressed by simply altering the cytoplasmic environment of a nucleus already becoming biochemically specialized. Further, it would seem that the repression is entirely physiological because the operated egg survives and the inhibited portion of the genome becomes synthetically active among the mitotic progeny in the cloned embryo with the appropriate sequence of RNA classes synthesized in a proper chronology.

Not only does egg cytoplasm have the capacity to modulate transcriptional activity of inserted nuclei (as revealed in *Xenopus*), it also alters patterns of DNA replication (as revealed in *Rana*). Blastula chromosomes incorporate ^{3}H-thymidine throughout their length. Chromosomes from tailbud embryos (Stage 17, Appendix VI) display terminal labeling. Blastulae produced from nuclei of tailbud stage cells, inserted into activated and enucleated eggs, display a pattern of DNA replication which mimics that of a normal blastula (Matsumoto and Dasgupta, 1974).

More Studies of Cytoplasmic Effects on Nuclei

Amphibian ova superficially appear very similar just before and just after ovulation. However, profound differences characterize ovarian ova (referred to as oocytes) and ovulated mature ova. Growth from a small unpigmented oocyte to a large, ready to be ovulated oocyte takes months. During the period of oocyte growth, there is no chromosome division but the oocyte is producing a variety of kinds of RNA (Figure 3-3). The oocyte nucleus, often termed the "germinal vesicle," is extremely large such that it can be seen with the unaided eye after manual dissection from the egg (Duryee, 1938; Battin, 1959). The enormous germinal vesicle approaches 1 mm diameter in some amphibians (Smith, 1912).

At the time of ovulation, which occurs spontaneously in nature, or as the result of hormone stimulation of intact amphibians (Appendix I) or in vitro (Dettlaff, Nikitina, and Stroeva, 1964; Subtelny, Smith, and Ecker, 1968; Brachet, Pays-de-Schutter, and Hubert, 1975) the enlarged germinal vesicle breaks down, a spindle forms with chromosomes, and the ovum prepares to extrude the first polar body in the meiotic division that results in the formation of a mature egg. Neither RNA nor DNA is produced at the time of the first maturation division. The climax of meiotic maturation is activation of the egg by sperm entry or by manipulation in the laboratory (Appendix I). Sperm that were formerly quiescent with respect to DNA synthesis begin DNA production within minutes of penetration. No RNA synthesis occurs at this time.

In contrast with bone marrow, skin, stem cells of the gut, and some other adult cell types, brain cells are virtually mitotically inactive. Probably less than 0.1% of adult brain cells divide. While the vast majority of brain cells in an adult frog do not replicate DNA, they do produce RNA. It is exciting indeed to contemplate that morphologically and biochemically specialized brain cell nuclei obtained from adult frogs will respond to cues from egg cytoplasm. Not only will they react to oocyte and mature egg cytoplasm but the response is appropriate. Consider the following: Adult brain nuclei, when inserted into mature egg cytoplasm, incorporate ^{3}H-thymidine. This means that they are producing new DNA. The brain nuclei, which normally are not producing DNA, start doing so perhaps in response to the same stimulus that invokes a sperm nucleus to produce DNA when it finds itself in mature egg cytoplasm. Appropriately, RNA synthesis is inhibited.

What is here described for adult brain nuclei of frogs is found to obtain also for adult liver and blood cell nuclei. Whatever it is in the mature egg cytoplasm of frogs that permits or evokes DNA replication of these several types of mature frog nuclei is equally effective in stimulating DNA replication in the nuclei of other species. Indeed, mouse liver nuclei synthesize DNA after injection into mature frog eggs.

It is not egg cytoplasm per se that starts the adult nuclei on a course of DNA synthesis but it is a special kind of cytoplasm — the cytoplasm of a mature egg. This is known because insertion of adult brain nuclei into growing oocyte cytoplasm results in *no* DNA synthesis. However, there is an enhancement of RNA synthesis. Further, as the oocyte matures, the adult brain nuclei lose their nuclear membranes and chromosomes appear on spindles (Graham, Arms, and Gurdon, 1966; Gurdon, 1967a, 1968a, see also Ziegler and Masui, 1973).

The oocyte is not only metabolically different from the mature egg but its nuclear morphology similarly is strikingly distinct. The germinal vesicle, as indicated above, is huge. Nuclei transplanted to oocyte cytoplasm mimic to a rather considerable extent the germinal vesicle. Blastula nuclei implanted into growing oocyte cytoplasm enlarge to over 250 times their original volume. Adult cell nuclei swell also but more modestly. Embryonic nuclei enlarge less than blastula nuclei but more than adult nuclei. Thus, the extent of nuclear enlargement is a rough measure of the extent of cell differentiation. In the course of these experiments, some nuclei were transplanted with intact plasma membranes. The nuclei of intact cells do not swell. The lack of response on the part of cells protected with an intact plasma membrane is reminiscent of failure of whole cells, when transplanted to enucleated mature eggs, to participate in embryonic development (see discussion in Appendix II). In summary, "the main outcome of these experiments is the demonstration that the nuclei of embryonic as well as differentiated cells quickly assume, in all respects investigated, the type of activity characteristic of growing and maturing oocytes into whose cytoplasm they have been inserted" (Gurdon, 1968a).

Recently, it has been shown that DNA synthesis in isolated nuclei can be induced by a cell-free cytoplasmic extract system. A protein(s) that appears to initiate DNA synthesis is found at high levels in eggs and early embryos but is less active in oocytes, older embryos, and adult tissues (Benbow and Ford, 1975).

Microinjection of Macromolecules into Ova of Xenopus

Cloning is a method of asexual reproduction. Microinjection of macromolecules is not cloning. However, the major rationale for performing nuclear transplantation experiments is to provide insight for the understanding of genomic interaction with cytoplasm. Certainly, one of the virtues of the test system is that it is natural. A nucleus is asked how it will perform in a milieu of normal and unmutilated cytoplasm. Normal frogs arise from many of these experiments. The frogs demonstrate that test nucleus and recipient cytoplasm are not irreparably damaged by the manipulations. A logical next step in the analysis of development is to ascertain if biologically significant macromolecules will persist, replicate, or function when injected into test eggs. If they do, then the characterization of the conditions under which the molecules persist, replicate, or function should be possible.

A recently ovulated egg deprived of a nucleus has cytoplasm capable of not only sustaining an implanted somatic nucleus as we have seen previously but the genetic material itself is replicated in a non-nuclear cytoplasmic system. There is synthesis of double-stranded DNA injected into eggs but none with oocytes. In contrast, single-stranded DNA replicates in both oocytes and mature eggs (Ford and Woodland, 1975).

Can an injected foreign messenger RNA (mRNA) be translated by egg cytoplasm into a gene produce that is detectable with conventional biochemical procedure? (Gurdon, Woodland, and Lingrel, 1974)? Messenger RNA obtained from immature red blood cells of rabbits was injected into *Xenopus* oocytes. The operated eggs produced rabbit globin but interestingly enough they did not produce *Xenopus* globin. The normal controls that inhibit production of a gene product until the appropriate ontological time remain functional despite the injection of the alien RNA. The foreign gene product in an otherwise normal oocyte is a cogent commentary on the normality of the reacting system. It would seem that translational machinery is available in excess of the normal needs of the oocyte because the cell continues in its normal activities in addition to making the novel gene product. It would seem that *any* message from a vertebrate can be translated in oocyte cytoplasm. Also, the controls regulating which gene is expressed seem not to be at the level of translation but at some level prior to translation.

Oocytes can be cultured in vitro for a number of days. Since the translation of a foreign message seems to be possible with messages from a number of different sources, one may seek to determine the stability of the inserted mRNA. Rabbit globin messenger was injected on day 1 of oocyte culture. Radioactive amino acids were injected on day 13. It was found that after almost 2 weeks the test system was still producing rabbit globin at 70% of its initial rate. What then is the half-life of the message? In this particular case, it must be exceedingly lengthy for at least some of the modest loss of activity may be attributed to the decline in viability of in vitro cultured oocytes (Gurdon, Lingrel, and Marbaix, 1973). Additional discussion concerning the control of gene activity as revealed with both intact nuclei and injected macromolecules may be found in Gurdon (1974b, c).

Cloning in Rana *and* Xenopus *Compared*

The microinjection studies of macromolecules are, of course, unique to *Xenopus* and provide new insight into mechanisms of gene activity. The cloning studies of *Xenopus*, however, do not appear to this author to be substantially different from those of *Rana*. Blastula nuclei of both species are totipotent and sexually mature frogs may be reared from nuclear transfer experiments. Nuclei from older embryonic donors become progressively less competent to participate in normal development as tested by nuclear transfer. This is equally true for *Xenopus* and *Rana*. Because the frogs are systematically quite far apart (i.e., they are not closely related anuran species), it is my judgment that the similarly reduced capacity of nuclei of older donors to promote normal development by the cloning procedure manifested by *Xenopus* and *Rana* reflects a fundamental change in nuclei that occurs with the differentiation process.

Urodele Nuclear Transplantation

The first successful cloning experiment utilized eggs and embryos of an anuran, *Rana pipiens* (Briggs and King, 1952; Chapter 2). Although there has been a modest information explosion relating to cloning, by far the majority of nuclear transplantation studies have utilized anurans. However, in the approximately quarter of a century that has followed those pioneering investigations of Briggs and King, a diversity of other kinds of amphibians have been exploited in cloning experiments. The other kinds of amphibians include assorted Ranids, toads, and tailed amphibia (Appendix IV). Perhaps even more remarkable, fish are now serving as a research resource for nuclear transplantation. The studies with alternative experimental animals are important because they demonstrate clearly that the basic procedure is adaptable to forms within both major orders of the Class Amphibia (Anura and Urodela) and also to an entirely different class of chordates, the Teleosts. The relative ease of extending the mechanics of cloning to such an array of chordates should give confidence of ultimate success to those individuals who seek to perform cloning in mammals (Bromhall, 1975). The nuclear transfer experiments with tailed amphibia are particularly interesting because they provide data which allow us to view with a perspective the more extensive information acquired from *R. pipiens* and *Xenopus laevis*.

There is, of course, another reason why an experimental embryologist might have substantial interest in nuclear transfer operations that utilize urodeles. Genetic variants of both *R. pipiens* and *X. laevis* are few in number (Appendix I). The axolotl (Figure 5-7), in vivid contrast, is rich, with 35 developmentally significant mutations (Briggs, 1973; Malacinski and Brothers, 1974). Thus, it should

Figure 5-7. Pigment mutants of the axolotl, *Ambystoma mexicanum*. Shown are white (*left*), yellow albino (*center*), and the dark wild-type (*right*). (G. M. Malacinski's cover illustration for *Science*, vol. 184, No. 4124, June 14, 1974.) Copyright 1974 by the American Association for the Advancement of Science.

be clear that a technology for cloning with newts and salamanders was urgently needed after the procedure proved feasible with frogs. Finally, recall that the first experiment with a vertebrate embryo that demonstrated nuclear equivalence was the unsegmented zygote constriction experiment of Hans Spemann on an urodelen form. Twin embryos were obtained by delayed nucleation of an egg fragment. Further, Spemann on the basis of his experience with urodele embryos anticipated nuclear transplantation (Spemann, 1938; Chapter 1). It seemed appropriate therefore that this classic material of experimental embryology would be ultimately exploited in the cloning process.

Nuclear Transfer in Tailed Amphibia: A History
That Precedes That of Frog Cloning

Probably the first modern attempt at nuclear transfer with a chordate egg was that of Professor Georgiĭ Viktorovich Lopashov, Institute of Developmental Biology, Soviet Academy of Sciences, Moscow, who grafted embryonic nuclei into unsegmented eggs of *Triton*. Blastulae formed but developed no further (Lopashov, 1945).

The initial failure with newt nuclear transfer to obtain an operated egg that developed beyond a blastula was followed in time by the well-known success with frogs by Briggs and King (1952). Perhaps the success of the investigators with frogs stimulated a new trial with newt eggs. Waddington and Pantelouris (1953) reported no better results, however. They had no method for enucleation of a newt egg, so they allowed a constricted zygote to undergo initial cleavage. The unsegmented fragment of the constricted egg was presumed to be devoid of a nucleus and accordingly became the recipient of a grafted donor nucleus. Segmentation but no real gastrulation occurred after insertion of nuclei obtained from blastulae, gastrulae, and neural plate stage *Triturus palmatus* donors. One might postulate that the developmental block occurred because of damage to donor nuclei. In other words, is it possible for embryonic nuclei of *Triturus* to be taken up in a micropipette and thrust into the cytoplasm of a recipient egg without trauma?

Clearly, the answer to that question is yes. Haploid nuclei were grafted to fertilized and non-enucleated recipient ova. The diploid host therefore could be distinguished cytologically from haploid donor. Both haploid and diploid nuclei were recognized in all three germ layers of the resulting chimaeras, which showed, among other things, that the manipulation of donor nuclei at transplantation does not necessarily result in lethality (Pantelouris and Jacob, 1958).

Another early attempt at nuclear transplantation with urodeles was that of Lehman (1955, 1957). He used whole ova (in contrast to the egg fragments of Waddington and Pantelouris, 1953) that were activated with X-irradiated sperm. The maternal genetic material was removed by pricking the egg spot with a sharp glass needle and sucking the maturation spindle out through the puncture with a mouth micropipette (Figure 5-8) after the method of Curry (1936). One would be inclined to believe that development of a whole egg would be superior to an egg fragment but, unfortunately, development blocked after cleavage of the nuclear transplant egg.

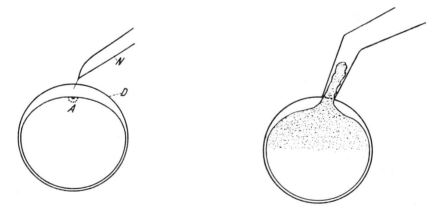

Figure 5-8. Method for removing maternal chromosomes from an unfertilized egg of tailed amphibian. A sharp glass needle is used to puncture the egg membrane (*left*) and the maternal chromosomes with spindle and some cytoplasm are withdrawn (*right*) by suction exerted by means of a mouth held micropipette. (From Curry, 1936.)

Success with Urodele Nuclear Transfer

Success came after the disappointing endeavors reported above. Jacques Signoret and his colleagues, working at Indiana University, succeeded in transplanting blastula nuclei of the Mexican axolotl, *Ambystoma mexicanum* (Signoret, Briggs, and Humphrey, 1962). Success was reported at essentially the same time utilizing the triton, *Pleurodeles waltlii* (Signoret and Picheral, 1962). Procedural differences varied from nuclear transfer experiments in frogs but the principles of the operation were the same (see Appendix II for discussion of heat shock or electric shock activation of the egg and ultraviolet ablation of maternal genetic material in urodele eggs). In both species, blastula nuclei were shown to be totipotent, i.e., adults were obtained after blastula nuclear implantation. Sexually mature adult axolotls produced from blastula nuclear transfers are illustrated in Brothers (1976).

Before proceeding further, it is well to emphasize that a genetic marker associated with the donor nucleus, which is subsequently expressed in the developing nuclear transplant, is essential to exclude the possibility of faulty enucleation of the recipient egg. It was stated above that the axolotl has a wealth of genetic variants. One of these, the "white" mutant is well studied (Dalton and Krassner, 1959). Pigmentation is controlled by the alleles D and d. Normally pigmented larvae result from Dd or DD genotypes, while the homozygous recessive dd results in "white." The experimental design was to insert nuclei carrying D into the enucleated eggs of dd females. All nuclear transplant larvae (37) in the first series of experiments were black and diploid (Figure 5-9), thus demonstrating that the implanted nucleus was the nucleus responsible for development (Signoret, Briggs, and Humphrey, 1962).

What is the capacity of nuclei of more mature embryos to promote embryonic differentiation? Neurulae served as nuclear donors in experiments with the axolotl. Nuclei of several cell types were transplanted. Each cell type investigated could promote cleavage. The capacity of neurula nuclei to promote cleavage varied from 53% in lateral plate mesoderm nuclei to 75% in endoderm nuclei. The ability to stimulate further development was severely restricted however. Thus, none of the complete blastulae

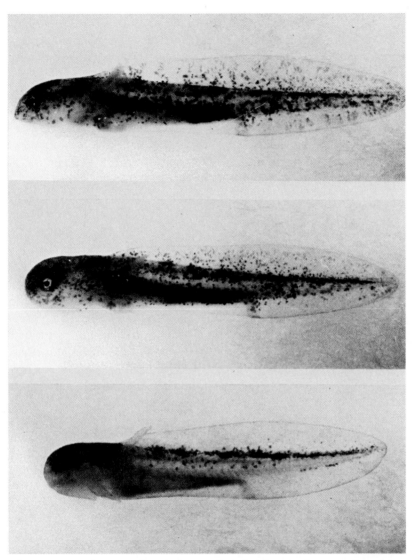

Figure 5-9. Nuclear transplant and control axolotl larvae. The upper larva has widely dispersed pigment cells characteristic of the DD or Dd genotype. It developed from a nucleus that contained the D allele injected into an enucleated ovum from a dd "white" female. Normally fertilized control larvae of DD genotype (*middle*) and dd genotype (*bottom*) may be compared with the nuclear transplant animal. Pigment cells are limited in distribution in the dd "white" larva.

(From Signoret, Briggs, and Humphrey, 1962.)

formed from neurula mesoderm transfers developed beyond abnormal gastrulation. One abnormal neurula was formed from among 31 blastulae developing from medullary plate nuclear transfer. Endoderm nuclei derived from neurulae performed less well than lateral plate mesodermal nuclei. One normal larva ensued from 362 blastulae formed from notochord nuclear transfer.

Cytological examination of the nuclear transfer embryos revealed mitotic abnormalities such as chromosome bridges and ring chromosomes. Abnormal urodele nuclear transfer embryos were thus correlated with chromosomal abnormalities much as they have been in studies of anuran nuclear transfer embryos (Chapter 3). The abnormalities in development could not be reduced by using different nuclear transfer solutions or by increasing the ploidy of the donor nucleus. Perhaps significantly, the region of notochord or endoderm (anterior versus posterior) did affect the yield of transplant embryos that developed beyond gastrulation. Since the difficulties in handling anterior notochord cells are no greater than those in handling posterior notochord cells, but anterior nuclei rarely promote development beyond cleavage in contrast to transplant embryos that develop to the tailbud stage from posterior nuclei, the authors suggest that the state of nuclear differentiation controls the yield of transplant embryos (Briggs, Signoret, and Humphrey, 1964).

Blastula, neurula, and tailbud nuclei of *Pleurodeles waltlii* seem equal in their capacity to promote blastula formation, but there is a decreased ability to promote gastrulation correlated with the increased age of the embryo providing donor nuclei. Although the yield of gastrulae produced by transplanting tailbud (Stages 22-23, Gallien and Durocher, 1957) endoderm nuclei is low, some of the gastrulae develop to the young adult stage (Picheral, 1962). Not surprisingly, and in harmony with the findings of both anuran and axolotl nuclear transplantation studies, some of the abnormal nuclear transplant *P. waltlii* were found to have an abnormal complement of chromosomes (Gallien, Picheral, and Lacroix, 1963b).

Discussions speculating as to the cause of chromosomal abnormalities found in urodele transplantation embryos are found in Briggs, Signoret, and Humphrey (1964), Signoret (1965), and Gallien (1974).

Nuclear transfer in *Triturus vulgaris* of blastula to early neurula nuclear donors was reported by Sládeček and Mazáková-Stefanová (1964). The report is disappointing because recipient eggs developed no further than free swimming tadpoles (recall that blastula nuclei were shown to be totipotent in both the axolotl and *P. waltlii*). Further, they implanted presumably diploid donor nuclei into irradiated, fertilized recipient ova. Failure of the radiation to eliminate the maternal chromosomes and inability of the inserted nucleus to participate in development would result in a diploid embryo that would be difficult to distinguish from a successful diploid nuclear transfer embryo. Nucleocytoplasmic hybrids produced by these workers (Sládeček and Mazáková-Štefanová, 1965; Sládeček and Romanovský, 1967) will be discussed in Chapter 7.

A Cytoplasmic Factor Controlling Post-Gastrular Development

The analysis of how the cytoplasm reprograms inserted nuclei to mimic zygote nuclei is now of greater importance than seeking an answer to whether or not somatic nuclei are developmentally totipotent. Thus, it is now known that nonhistone proteins move bidirectionally between an inserted endoderm nucleus and the egg cytoplasm (DiBerardino and Hoffner, 1975; Hoffner and DiBerardino, in press). It is also known that an egg cytoplasmic extract can induce DNA synthesis in isolated nuclei (Benbow and Ford, 1975). Therefore, the recent study of Brothers (1976) is of substantial interest to developmental biologists. Brothers has utilized one of the many axolotl mutants referred to earlier. Females that are homozygous recessive for *o* (for ova deficient) produce eggs that are unable to develop beyond gastrulation. Zygotes from such females will develop normally if they receive an injection of germinal vesicle material or mature egg cytoplasm from wild-type ova (Briggs and Cassens, 1966). Analysis by nuclear transfer demonstrates the presence of a cytoplasmic factor (a protein deficient in the *o* mutant) that is necessary for activation of blastula nuclei to participate in subsequent development. The study of Brothers is a model that may permit the isolation and characterization of a cytoplasmic substance essential for selective gene expression.

Regeneration Blastema Nuclear Transfer

Regeneration of missing parts has long been a favorite subject of study for embryologists (Rose, 1970). While it is true that *Xenopus* retains substantial capacity for "hypotypic" regeneration (Goode, 1967), and preliminary observations have been reported concerning very limited development promoted by *Xenopus* blastema nuclear transplantation (Burgess, 1967; 1974), newts and salamanders have been the classic material for limb regeneration studies.

Cells which appear to be morphologically dedifferentiated appear in about 1 week distal to the cut surface of the *Ambystoma* amputated forelimb (Hay, 1959). The dedifferentiated cells, which are known as blastema cells, persist in the regenerating stump for about a week. Muscle, skin, cartilage, etc., become histologically recognizable in the blastema after this time. There is an emerging body of evidence that strongly suggests that the blastema cells that appear to be dedifferentiated can, in fact, give rise to other kinds of differentiated cell types (Namenwirth, 1974; Steen, 1968). Do the cells of a blastema, which appear to be embryonic both morphologically and in the capacity to give rise to alternate types of differentiated cells, behave as embryonic nuclei when tested by nuclear transplantation? One should recall that the axolotl exhibits severe restriction in the capacity of post-blastular nuclei to promote normal development when tested by cloning. The etiology of the early restriction is not known (see above). The capacity of blastema nuclei from advanced larvae to promote the development of an enucleated egg is therefore of great interest.

The front limbs of *Ambystoma* larvae 4 to 6 inches in length were amputated. Nuclei obtained from blastemas which formed at the site of amputation were tested by nuclear transplantation. Segmentation occurred in activated and enucleated eggs after insertion of nuclei obtained from blastemas 7 to 16 days after amputation. One egg, which received a nucleus from an 8-day regenerate, gastrulated and formed an abnormal tailbud embryo. An additional abnormal tailbud embryo formed in another experiment (Dasgupta, 1970). No genetic test of the transplanted blastema nucleus was described that would authenticate the validity of the study.

However, in an earlier series, a fragment of a nuclear transfer blastula produced from a transplanted blastema nucleus of the genetically white strain (dd) was grafted to a dark embryo (DD). The dark host had a patch of white tissue which suggested that the implanted blastema nucleus was indeed the genetic material responsible for development (Dasgupta, 1969; 1970).

The blastema nuclear transfer animals were aneuploid, which probably accounts for their abnormal development. The genesis of the aneuploidy, as in other nuclear transfer animals, is unknown.

The blastema nuclear transfer studies should be continued because they seem to be one of a class of nuclear transfer studies in which nuclei carrying out their customary physiological activities do not interact with egg cytoplasm to form an embryo, but when the donor cells are provoked into mitotic activity and novel physiological tasks, they become competent to promote at least limited embryonic development. Although normal nuclei from larval limbs of the same age as the experimental blastema nuclei were not transplanted in the study of Dasgupta, it was already known that totipotency was limited to nuclei of blastula (in axolotl). One could surmise, therefore, that nuclei obtained from ordinary limbs of larvae 4 to 6 inches long could not promote embryonic development. Yet, blastema nuclei resulted in two tailbud embryos after nuclear transplantation. These studies may be compared with nuclear transfer of tumors (Chapter 4) and cultured skin of adult *Xenopus* (Chapter 6). There seems to be an awakening of the capacity to program for alternate cell types associated with enhanced mitotic rate that occurs as a blastema forms, as cells become neoplastic, or when cells are cultured in vitro.

Cloning Studies Made by Investigators in the USSR

Stroeva and Nikitina (1960) demonstrated clearly that many nuclei of late blastulae and early gastrulae of *R. temporaria* and *R. arvalis* promote cleavage upon transplantation. Further, some of the nuclear transfer blastulae will gastrulate and a few will develop to the hatching stage. These experiments unequivocally established the capacity of late blastula and early gastrula nuclei of the two species to participate in substantial embryonic development.

In later studies, Nikitina (1964, 1969) reported a decreasing yield of cloned embryos that develop associated with an increase in the age of the nuclear donor embryo. She noted that while late blastula and early gastrula nuclear transfers in *Bufo viridis* will provide a large yield of tailbud embryos (41% of transplant blastulae form tailbud embryos), tailbud nuclei when transplanted will produce less than half as many embryos that develop to the tailbud stage (20% of transplant blastulae form tailbud embryos in the latter case). However, normal tadpoles were formed from transplanted eyecup ectoderm nuclei from tailbud embryos in a few cases. The tadpoles lost their gills, developed hind legs, but died prior to metamorphosis. Decreasing yields of successful nuclear transfers were also recorded with the implantation of progressively older nuclei obtained from *R. arvalis*, *R. temporaria*, and *B. bufo* (Table 1).

Table 1. Development of Enucleated Eggs after Diploid Nuclei
Were Transplanted to Them

Species	Experimental Series No. [a]	No. of Operated Eggs	Stages of Transplant-Embryo Development			
			Blastula	Gastrula	Tailbud	Tadpole
Bufo viridis	I	140	41(29%)	27(66%)	20(49%)	8
	II	117	24(21%)	13(54%)	9(38%)	5
	III	106	19(18%)	9(47%)	5(26%)	2
	IV	110	18(16%)	6(33%)	3(17%)	1
Bufo bufo	I	139	34(24%)	22(65%)	14(41%)	7
	II	118	23(20%)	10(43%)	8(35%)	3
	III	76	13(17%)	5(39%)	2(15%)	1
	IV	57	9(16%)	4(44%)	2(22%)	1
Rana arvalis	I	285	51(18%)	22(43%)	13(26%)	3
	II	46	7(15%)	2(29%)	1	—
	III	64	7(11%)	2(29%)	1	1
	IV	91	8(9%)	2(25%)	1	1
Rana temporaria	I	267	27(10%)	10(37%)	5(19%)	2
	III	91	6(7%)	2(33%)	1	1
	IV	91	7(8%)	2(29%)	1	1

Source: Adapted from Nikitina, 1969.

[a]Series I are cloning experiments with donor ectoderm nuclei from blastulae or early gastrulae; Series II have donor nuclei from ectoderm of late gastrulae; Series III have donor nuclei from the eye anlage from the early neurula stage; and Series IV have donor nuclei obtained from the eyecup of tail-bud stage embryos.

I would like to suggest that the most significant contribution of the studies of nuclear implantation in anurans by the Soviet scientists is that they reinforce the conclusions that have been derived from the studies in *Rana pipiens* and *Xenopus laevis*. These conclusions are that blastula nuclei are developmentally totipotent and that the ability to promote normal development upon transplantation diminishes rapidly with increased age of the donor nucleus. There are variations in the embryonic stage and tissue type associated with loss of nuclear potencies as measured by nuclear transfer, but inexorably the potencies are repressed. Chromosomal abnormalities are associated with decreased developmental potentialities but the genesis of the abnormalities in the cloned embryo is yet to be ascertained.

More recently, the Russian scientists have become interested in the maturation of oocytes and the role of the germinal vesicle in maturation. They have made nuclear transplantations of embryonic nuclei and oocyte nuclei as a means of testing the state of egg maturation (Dettlaff, Nikitina, and Stroeva, 1964; Nikitina, 1972, 1974).

Nuclear Transplantation Studies in China

The first Chinese nuclear transplantation studies that I know about took place 11 years after those of Briggs and King (1952). No review of these studies has been included in any western paper but western studies are certainly known to the Chinese (Tung and Tung, 1963). The Chinese papers will be discussed here because the Chinese pioneered (and are thus far unique) in studies of cloned fish and because the published results of amphibian nuclear transfer from China are not entirely in harmony with the cloning studies of either *Rana* or *Xenopus*. Accordingly, they merit scrutiny.

Professor T. C. Tung of the Institute of Zoology, Academia Sinica, Peking, China, described his nuclear transplantation procedure in the Chinese language originally in 1963 (Tung et al., 1963). It was subsequently published again in English (Tung et al., 1965; see also Tung et al., 1973). They were the first to report cloning in fish. They used two species, viz., the wild gold fish "chi-yü," *Ca-*

rassius auratus, and a carp, *Rhodeus sinensis*. The fish were chosen because of their ease of rearing in the laboratory and because one, *C. auratus*, reaches sexual maturity in one year. Enucleation of recipient eggs was accomplished manually with the aid of a glass microneedle. Dissociated donor cells were drawn up into a micropipette by means of a micrometer syringe (manufactured in Shanghai, China) and inserted into the recipient egg. The micromanipulator was the mechanical stage of a compound microscope mounted vertically. Nuclear transplant fish developed in some cases to the swimming stage ("fry").

The availability of a cloning capability shortly led to an analysis of embryonic nuclei of different stages in an amphibian. The amphibian chosen for study was the toad, *Bufo bufo gargarizans*. The eggs of this species are easily ovulated (Wang, 1963) and the nuclear transfer procedure, previously described for fish, was adapted to the new form. The Russian nuclear transplanters had previously used a different subspecies, *B. bufo asiaticus*. Tung et al., (1964) reported on the developmental capacity of endoderm nuclei derived from embryos of gastrulae to hatched tadpoles (Table 2). It may be seen by reference to the table that only a few nuclear transfer blastulae develop to the swimming tadpole stage. Thus, 3% of blastulae formed from the transfer of gastrula endoderm nuclei

Table 2. Development of Endoderm Nuclear Transplant
Embryos of *Bufo bufo gargarizans*

Donor Embryo Age Group	Stages of Transplant-Embryo Development						
	Blastula	Gastrula	Neurula	Tailbud Embryo	Hatched Tadpole	Swimming Tadpole	Adult Toad
Blastula	392	193	89	58	29	11	2
(control)	100%	49%	23%	15%	7%	3%	0.5%
Gastrula	212	107	60	20	10	7	2
	100%	50%	28%	9%	5%	3%	1%
Neurula	194	91	34	24	20	11	2
	100%	47%	18%	12%	10%	5%	1%
Tail-bud	55	12	7	1	1	1	
embryo	100%	22%	13%	2%	2%	2%	
Hatched	173	71	29	15	13	7	
tadpole	100%	41%	17%	9%	8%	4%	

Source: Tung et al., 1964.

develop to the free swimming tadpole stage. Surprisingly, the yield of nuclear transfer embryos that develop to the swimming tadpole stage is independent of the age of nuclear donor, at least with nuclei obtained from gastrulae to hatched tadpole stages. Tung and his coworkers reported two adult toads derived from transplanted endoderm nuclei of neurulae.

A similar study was made by S. M. Tung (1964) concerning the developmental capacity of mesodermal nuclei of *B. bufo gargarizans*. It was reported that mesoderm nuclei competent to promote cleavage decrease as the age of the donor becomes older. However, the fraction of nuclear transplant blastulae that will develop to the swimming tadpole stage remains at about one-third. One adult toad was produced by the transfer of a mesodermal nucleus obtained from a tadpole of the muscular response stage. No "mesodermal syndrome" comparable to the endoderm syndrome (Chapter 3) was described.

If it can be assumed that the decline in nuclear transfer success in producing blastulae associated with increased age of nuclear donor is due to technical difficulties and the true nature of the differentiated donor nucleus is revealed by post-blastular development, then it would seem that the nuclear transfer procedure either reveals that *Bufo bufo gargarizans* nuclei do not continue to differentiate with age, or that whatever nuclear differentiation there is is reversed by the procedure. It should be remembered that these results are at variance with the outcome of experiments with *Rana* and *Xenopus*, which are certainly more extensive. Further, chromosome analysis and the use of cytogenetic markers are not reported. Note also that in contrast with *R. pipiens* but similar to *X. laevis*, no germ layer-specific syndrome of abnormalities was reported. Hence, for the time being, it would seem that the studies with *Bufo bufo gargarizans* are best considered as interesting exceptions to some of the conclusions drawn from studies of *R. pipiens* and *X. laevis*.

More recently, Tung and his associates studied the interaction of transplanted embryonic nuclei with highly specialized egg cytoplasm of the ascidian *Ciona*. Cytoplasmic control of nuclear function is elegantly illustrated with this unique system (Tung, et al., 1977).

CHAPTER 6

The Cloning of Adults and Other Transfers of Mature Nuclei

Nuclear transplantation studies of adult nuclei certainly are far fewer in number than cloning studies of embryonic nuclei. This is curious indeed. Curious because if one is aware at all of media response to the craft, one notes that cloning of adults is where all the action is, or where it is supposed to be.

This chapter will present the meager studies known to the author concerning the cloning of adult nuclei. I suppose that to some the sparseness of nuclear transfer studies of adults will come as a disappointment, considering all that has been written about them. I would like to suggest that an appropriate response to the literature is not disappointment but it should be pleasure and anticipation. Pleasure would be appropriate because if adult totipotency had been shown to exist by the cloning procedure back in the 1950s in those early days of nuclear transplantation, then I suspect that the technique would have been abandoned long ago as a near worthless technical tour de force that would demonstrate more about one's manual dexterity than about cell biology. For, totipotency of adult nuclei would simply mean that the cytoplasm of eggs could completely and at once reverse specializations that have been imposed on nuclei during the course of differentiation to the adult stage. Prior to and emerging with the cloning era were concepts of genetic equivalence of somatic nuclei. Thus, it came as no shock to biologists that blastula nuclei, when transplanted to enucleated eggs, were shown to be totipotent. It was anticipated at the time that as technical skills developed adult nuclei would also be shown to be totipotent by the same procedure. This did not occur. How pleasant. Pleasant because if adult nuclei were shown to be totipotent by the same procedure that blastula nuclei were shown to be totipotent, nuclear transplantation would not be a tool for the further study of the regulation of gene expression. Adult cells are indeed different than embryonic cells. If the cloning procedure did not reveal this, then it would be worthless for the study of differentiation. But, profound limitations were soon noted in the capacity for most embryonic nuclei to promote normal development. This limitation provides for a measure of rational anticipation. I think that we can anticipate that with ingenuity and labor we will develop new methods that will allow for the unfolding and expression of the entire genome of differentiated cells. Already there is enticing evidence that this has or is about to occur. That evidence will be presented below.

It is important to note that because the adult nucleus is *not capable* of immediate expression of its latent genetic potentialities when thrust into an enucleated egg, manipulations of that adult nucleus become necessary to coax expression of those potentialities. I believe that much can be learned from a careful analysis of the nature of the manipulations. If several kinds of perturbations result in highly differentiated mature cells of different sorts becoming like zygotes with respect to their capacity for development, then analysis of common factors, if any, that ensue from the perturbations may yield new insight into how cells become different. I have faith that this will be an end result of cloning studies.

If cloning makes any contribution in the late twentieth century, it will *not* be the replication of new individuals of political or scientific note, as is so often depicted by the media, but it will be a better understanding of how a common genome comes to be expressed in one environment as liver and in another environment as brain. It is not a different genome that results in liver being dissimilar from brain but the differences in liver and brain relate to selective expression of parts of a common genome. The cloning procedure applied to adult nuclei abruptly changes the environment of the genome, but at the present time mature egg cytoplasm by itself seems to be an inadequate environmental stimulation for eliciting totipotency. What other environments are needed to reveal an intact genome?

As we learn more and more about closely related species, it seems that cellular and molecular similarities between the species are often more profound than dissimiliarities. Until very recently, it was difficult if not impossible to tell one small human chromosome from another. Within the past few years, however, a great diversity of banding techniques has been developed. The techniques allow for the precise identification of each autosome of man (Caspersson, Zech, and Johansson, 1970; Comings and Avelino, 1975). More important for this discussion, the procedures allow for comparison of human chromosome with chromosomes of closely related species such as chimpanzees. While anatomically and behaviorally chimpanzees and humans vary, there is an incredible similarity between their chromosomes as revealed by banding (Bobrow and Madan, 1973). Perhaps even more important, comparison of a large number of human and chimpanzee proteins reveals that the proteins are more than 99% alike (King, 1974). These observations have suggested that mutations affecting the *expression* of genes rather than mutations affecting structural genes may be critically important in evolution, especially in humans. These views are supported by interspecific hybridization studies related to investigations of similarities between serum albumins of a number of species. There is a close molecular similarity of serum albumins between mammalian species that are competent to hybridize, which suggests to some that regulatory evolution has occurred more rapidly than protein evolution (Wilson, Maxson, and Sarich, 1974a; see also Wilson, Maxson, and Sarich, 1974b). Thus, understanding *regulation* of gene expression in higher organisms may give considerable insight into the origin of new species (Kolata, 1975).

It is not the purpose of this chapter to discourse on Darwinism but I would like to hope that there may be an even broader application or value of cloning studies than previously thought. Cloning studies may be relevant not only to ontogeny but they may become useful to studies of phylogeny. Certainly it is anticipated that nuclear transplantation studies will yield information about regulation or control of gene expression. Thus, cloning may become as important to general biology as it is thought now to be important to cell biology and embryology. That importance will develop from studies of nuclei that are not at first sight totipotent but which become so because of physical and chemical manipulations of their environments—those manipulated nuclei are of course obtained from cells of maturing and adult organisms. Cloning of those nuclei is the subject of this chapter.

What Is an Adult Cell?

There is a simplistic answer to the question posed in the section heading. An adult cell is a cell obtained from a fully mature individual. However, adults produce gametes as a rule. Is a mature egg or sperm an adult cell? For most embryologists, gametes are highly specialized but genetically undifferentiated cells. Thus, while obtained from a non-juvenile, an egg is not an adult cell. To minimize confusion, perhaps the question should be worded so as to eliminate germ cells. Rephrased, it would read "what is an adult somatic cell?" Some might answer that an adult somatic cell is a nongerminal cell derived from a *mature* organism, and it is frequently assumed to be equivalent to a "fully differentiated cell." The discourse can become bewildering at this point because perfectly competent biologists have referred to "undifferentiated" cells of adults and others have referred to "fully differentiated cells" of embryos. Examples of certainly less than fully differentiated cells found in the adult include the cells of the skin stratum germativum (Wier, Fukuyama, and Epstein, 1971), blood stem cells (Wolf and Trentin, 1968), and glandular cells of the uterine endometrium during the proliferation phase (Cavazos et al., 1967). In contrast to adult cells that continue with the differentiative process are cells of embryos that are differentiated in form and function. Gurdon (1962a) referred to gut cells of embryos, many of which were less than a week old, as "fully differentiated." Cells between the digits of feet and forelimbs are programmed to become senescent and to die with the result that digits become free of the webbed state during early embryonic life (Saunders and

Fallon, 1966). Degeneration of neuroblasts occurs during the first week of incubation of the chick embryo (Hamburger and Levi-Montalcini, 1949). A "death clock" seems to be built into mesoderm cells of the chick embryo wing bud. The senescent cells are terminally differentiated and die at about 96 hours of incubation (Saunders, Gasseling, and Saunders, 1962). One might believe that cells that are programmed to die during ontogeny are performing a fully differentiated function by dying. Thus, adults have cells that are far from fully differentiated and embryos have cells that are so terminally differentiated that they meet an early demise.

What has all this got to do with a chapter on the cloning of adult cells? The comments are placed here to remind the reader that what is being studied in many nuclear transfer experiments is the process of differentiation. We seek answers to questions that relate to the reversibility or stability of the differentiated state. It is well to recall that DiBerardino and Hoffner (1971) suggested that it is more reasonable to relate developmental restrictions of transplanted nuclei to the state of differentiation of the donor cell than to the stage of development of the donor embryo. I agree.

Frogs Are Reported from Cloned Adult Lens Nuclei— Has Cloning of Adults Become a Fait Accompli?

Several years ago, Audrey L. Muggleton-Harris became interested in aging phenomena. She was able to transform a potentially "immortal" strain of amoebae to one with a fixed life span by limiting the food available to cultures (Muggleton and Danielli, 1958). Does the imposed limitation to life span of the altered amoebae have its roots in a changed nucleus or in transformed cytoplasm? An obvious mode for seeking an answer to the question is nuclear transplantation and, of course, nuclear transfer technology in amoebae had an early origin (Chapter 1). Nuclear transplantation suggested that the loss of immortality of the cultured amoebae had a complex etiology and it was due probably to changes in both nucleus and cytoplasm (Muggleton and Danielli, 1968).

Muggleton-Harris next turned her attention to a chordate that has an organ that is histologically pure and contains cells that vary from young to old. The chordate was *Rana pipiens* and the organ was the lens of the eye. Chordate lenses are not contaminated with vascular elements, nerves, or fibroblasts. The lens has a simple cuboidal epithelium at its anterior pole. The lens epithelium is mitotically active away from the pole and these dividing cells give rise to terminally differentiated elongate lens fibers which lack nuclei. It would seem, therefore, that the lens would be an ideal structure for assay of differentiation-related changes in nuclear potentialities.

Lens nuclei from pre-feeding embryos (Stages 22, 23, and 25, Appendix VI) were competent to support development only to gastrulation when inserted into activated and enucleated eggs (Muggleton-Harris, 1970). Adult frog lens cells will become mitotically active and form tumorlike growths after injury (Rafferty, 1963). When lens cells from adults were forced to become mitotically active by a pin prick, it was found that their nuclei, when inserted into enucleated eggs, would support development at least to the blastula stage (Muggleton-Harris, 1971a, 1971b). Another means of obtaining mitotically active lens cells is by organ culture (Gierthy and Rothstein, 1971). When nuclei from organ-cultured lens were transplanted, advanced embryonic (Muggleton-Harris, 1972) and *metamorphosed frogs* (Muggleton-Harris and Pezella, 1972) were reported. The nuclear transplant frogs were obtained not in the primary nuclear transfer generation but in the first serial transfer, i.e., by the transplantation of blastula nuclei obtained in the primary transfer. This to my knowledge is unique in the annals of *Rana pipiens* nuclear transplantation. No other investigator has reported totipotency of nuclei from adult tissue of the leopard frog.

What is the significance of this unique study? First, it should be noted that totipotency was not demonstrated in either embryonic or adult nuclei when transplanted directly. However, when the adult nuclei became mitotically active in organ culture, and when the nuclei were transplanted serially, a few frogs were obtained. Further, spermine (Hennen, 1970) was added to the nuclear transfer medium. Thus, an adult somatic nucleus may have been provoked by several very simple maneuvers to become totipotent. The maneuvers relate primarily to establishment of mitotic activity in adult tissue and perhaps the binding of a polycationic substance, spermine, to the DNA of the donor cell.

The cell type that was transplanted was known. Muggleton-Harris chose her donor tissue well, because lens is not contaminated with other cell types. "Adultness" was also clear. However, she did not reprogram terminally differentiated cells. Terminally differentiated, or "fully differentiated," lens cells become elongate and transparent and their nuclei die. What she transplanted were lens epithelial cells that were stimulated to divide before transplantation. They were not morphologically or functionally lens fibers.

Before proceeding further, a modest note of caution is in order. Parthenogenesis, however unlikely, can result in transformed frogs from activated eggs. It would seem to me that since the lens cloning study stands alone, i.e., it has not been repeated in other laboratories yet, nor have studies appeared from her laboratory subsequent to 1972, it would be useful for Muggleton-Harris to produce some triploid frogs (Briggs, 1947; Dasgupta, 1962), rear them to maturity, explant their lenses to a culture system, and utilize the triploid lens nuclei as genetically tagged nuclei. Cloned triploid frogs would be convincing evidence that the lens nuclei were truly reprogrammed. Since a few cloned lens frogs were reported to develop to metamorphosis, she could bypass the rearing of triploids altogether by purchase of adult Burnsi or Kandiyohi frogs (McKinnell and Dapkus, 1973) and utilize the pigment pattern mutants for genetic markers of the lens nucleus equally well (McKinnell, 1960, 1962b; Simpson and McKinnell, 1964). This modest note has been inserted because no data concerning timing of egg cleavage or efficacy of enucleation was included in the published reports. I, and I suspect many others, will be delighted to see lens cloning studies with nuclear markers. I would also be pleased to see a photograph of a "lens frog" and know whether or not such frogs can be reared to sexual maturity.

Lens and Tumor Cloning Compared

Have other adult somatic cells of *Rana pipiens* been transplanted? Yes. In Chapter 4 transfer of renal adenocarcinoma nuclei was described. The reader may notice some similarity between the studies. Transplantation of normal adult kidney nuclei results in little development of the recipient egg. One blastula formed from 31 nuclear transfer operations. This blastula served as nuclear donor for a second generation (serial) nuclear transfer experiment. Seven additional blastulae were formed but only one developed to the neurula stage — no others developed beyond gastrulation (King and DiBerardino, 1965). Exposure of frog kidney to a specific herpesvirus (Chapter 4) results in neoplasia, which means, among other things, that the kidney becomes mitotically active. The proliferating neoplastic kidney nuclei program for development to tadpole stages when inserted into activated and enucleated eggs. The tumor study is thus similar to the lens story in the sense that mitotically active adult nuclei promote· embryonic development, but it differs in the obvious fact that tumor nuclei, when transplanted, program for limited development thus far.

There is another difference. The tumor nuclear transfers were accomplished in both Philadelphia and in my laboratories in New Orleans and Minnesota. The lens nuclear transplantation studies are yet to be confirmed.

Adults Have Germ Cells That Are Genetically Totipotent — Are They Developmentally Totipotent?

Smith (1965, Chapter 3) reported that primordial germ cell nuclei of pre-feeding embryos (Stage 25, Appendix VI) were totipotent when tested by nuclear transfer. These cells that were totipotent contrasted markedly with the lack of normal nuclear transplant embryos obtained from gut nuclei. Thus, an embryonic stage may contain cells that exhibit both limited and total development as nuclear transplants. Does developmental totipotency persist in cells of the germ line, which may be thought of as genetically totipotent?

Blastula formation but no normal larvae were formed from the transfer of male germ nuclei from post-metamorphic males (DiBerardino, King, and Bohl, 1966). Later however, with improved microscopic technique, one larva was produced by the insertion of a spermatogonial nucleus into an enucleated egg (DiBerardino and Hoffner, 1971).

First, one may question what kind of cells are found in the testes of adult frogs. Fragments of dissociated testes contain blood cells, connective tissue, Sertoli cells, and spermatogonia. Spermatogonial cells may be recognized and selected for transfer because of size. Sertoli cells are larger and all other cells are smaller than the approximately 20 μ spermatogonia. The improved microscopic technique referred to above consists of an inverted compound microscope, a condenser and light source that can be easily elevated, and a dissecting microscope that can be moved into position above the inverted microscope in place of the elevated condenser and light source (see DiBerardino and Hoffner, 1971, figure 1).

The improved microscopic technique permits visualization of dissociated cells with a compound microscope, measurement of the cells with an ocular micrometer calibrated in micrometers, and visualization of the pickup of donor cells with the compound microscope. Transplantation of the broken donor cells into recipient eggs, however, is viewed with a conventional dissecting microscope.

Four embryos that developed beyond the neurula stage were reported. One began feeding. It lived for 20 days. It was formed from a transplanted spermatogonial nucleus derived from an *adult*. No feeding larvae were obtained from cloned spermatogonia obtained from juvenile frogs. All of the abnormal embryos were characterized by abnormal chromosome constitutions with the exception of the feeding larva, which was tetraploid and euploid. It is worthy of note that the most advanced development reported was a cloned spermatogonial cell derived not from a juvenile but from an adult frog (DiBerardino and Hoffner, 1971).

Mature Xenopus *Cells Cloned*

Nuclear transplantation of "fully differentiated" (Gurdon, 1962a) cells from tadpoles will not be discussed here because the donor nuclei were immature (see Chapter 5). More recently, cells from adult frogs have served as donors for nuclear transplantation experiments in *Xenopus* (Laskey and Gurdon, 1970). The adult cells were first placed in culture to obtain a population of rapidly dividing cells (Gurdon and Laskey, 1970a). Nuclei of cultured kidney, lung, heart, testes and skin were transplanted and these experiments were compared with transfers of Stage 40 (Nieuwkoop and Faber, 1967) cultured embryo nuclei. Most recipient eggs, receiving in vitro cultured adult nuclei, fail to cleave. However, some do form partial blastulae and these imperfect blastulae contain nuclei with substantial developmental potential. Serial transfer, i.e., transfer of nuclei obtained from partial blastulae produced by inserting nuclei of cultured cells into eggs, is a means of revealing the latent genetic potential of the original adult nuclear donor. Serial transfer experiments indicated that the several adult tissues listed above have the capacity to program for development to the heart beat stage (which occurs normally after about 1 day of growth). Further, all of the adult cell types except cultured heart could program for development to a swimming stage (which occurs normally during the third day of growth). These swimming tadpoles differ from normal tadpoles because they died, possibly from the effects of anal and cardiac edema (Lasky and Gurdon, 1970).

What is the differentiated state of skin nuclei that are cultured? Skin is but one of many adult tissues with cells in varying states of differentiation as discussed earlier in this chapter. In vitro cultures of skin used for transfers in *Xenopus* are believed to be "committed" (determined?) because after about a week the cells become filled with a birefringent material resembling keratin. The substance is inhibited by vitamin A and accelerated in synthesis by citral; a situation considered to be diagnostic of keratinization (Aydelotte, 1963). Donor nuclei for nuclear transfer were obtained *prior* to the filling up of cells with the keratinlike material. Are these adult cells, therefore, fully differentiated?

An answer to that question came several years later. Nuclei, obtained from skin cell culture that had been characterized with respect to its keratinization (Reeves and Laskey, 1975), were inserted into ultraviolet irradiated host ova. Over 99.9% of the donor cells contained keratin. Some of the recipient ova formed partial blastulae. These blastulae served as donors for retransplantation. Eight of the serially transferred partial blastulae formed clones of swimming tadpoles with well-differentiated tissues (Gurdon, Laskey, and Reeves, 1975). The most advanced tadpoles reached a developmental stage approximately that of Nieuwkoop and Faber (1967) Stage 41 or 42 at which time they became edematous and died without feeding. The tadpoles were produced with appropriate genetic markers associated with the donor nuclei (the nucleolar marker) and chromosome counts were made to assure that the development reported could be attributed to a nucleus from a keratinized skin cell.

The experiments with *Xenopus* may have something in common with the lens nuclei experiments. First, it should be noted that adult nuclei of *Xenopus* with no pretreatment do *not* promote embryonic development (Graham, Arms, and Gurdon, 1966; Gurdon, 1968a). Neither do they in *Rana* lens experiments. Second, if adult *Xenopus* cells become mitotically active in an in vitro culture system, they will form blastulae or partial blastulae when cloned. The fate of the primary transfer generation seems to be death but if serial transfer is undertaken, substantial embryonic development is possible in both *Xenopus* and *Rana*.

Several years ago I wrote about what seemed to me to be incredible similarities between cloning studies in *Rana* and *Xenopus* (McKinnell, 1972). I am still impressed with their likeness.

The Transfer of Adult Lymphocyte Nuclei

Initially, it was adequate in nuclear transfer studies to specify the stage or age of donor nuclei. Now, cytological or biochemical characterization indicative of the differentiated state of the donor cell population is essential for interpretation of nuclear transfer experiments. As examples, we know how dissociated frog renal tumor cells prepared for nuclear transfer fluoresce in the ultraviolet after treatment with acridine orange (Chapter 4) and we know that cultured skin cells used in cloning produce a keratinlike substance (see above). In a similar fashion, lymphocytes for nuclear transfer from *Xenopus* may be characterized as to their differentiated state (Wabl, Brun, and DuPasquier, 1975; DuPasquier and Wabl, 1977). Splenic lymphocytes from an immunized *Xenopus* were coupled by means of their immunoglobulin receptors to a nylon grid reacted with the antigen used for immunization. The lymphocytes, which were broken as they were drawn into a micro-pipette from the grid, were inserted into enucleated ova. Nuclear transfer blastulae and gastrulae were used as nuclear donors for serial transplantation. Tadpoles of one of the serial clones developed to Stages 43-44 (Nieuwkoop and Faber, 1967). Were the cells transplanted in fact lymphocytes? The authors reckon on the basis of inhibition of binding to the nylon grids by treatment of the cells with anti-immunoglobulin serum (and other observations) that 98% of the cells bore immunoglobulins and were therefore lymphocytes.

The lymphocyte cloning study is of special technical interest because the investigators were concerned about enucleation with ultraviolet irradiation. They reported that the enucleation method devised by Gurdon (1960a) could effectively eliminate the maternal pronucleus in as high as 89% of treated eggs (in good experiments) or as low as 51% (in other experiments). The reason for the variation in enucleation efficiency was not known but it emphasizes the *absolute* need for a *Xenopus* nuclear marker associated with the donor nucleus. McAvoy, Dixon, and Marshall (1975) also report imperfect enucleation with the ultraviolet irradiation procedure. The rationale for nuclear markers is discussed in Chapter 2.

Other Studies from Adult or Maturing
Xenopus *Nuclear Donors*

The effect of mitotic activity and stage in the cell cycle on the transplantability of gut nuclei was studied by McAvoy and Dixon (1974) and McAvoy, Dixon, and Marshall (1975). Trough cells, located between folds in the gut wall, are structurally unspecialized and divide asynchronously. They are thus in different phases of the cell cycle. Crest cells, in contrast, are mitotically inactive, are all in the same phase of the cell cycle, and are structurally specialized. Nuclear transplantation results were the same regardless of the kind of adult gut cell nucleus transplanted. Only a minority of nuclear transplants (0.4%) were capable of developing to the hatched tadpole stage and *none became frogs*. Serial nuclear transfer did not enhance embryogenesis among the cloned embryos.

Totipotent nuclei are found in skin of hatching tadpoles, which were over 2 days of age (Brun and Kobel, 1972). Even with such a young embryo as nuclear donor, development beyond gastrulation did not occur except by means of serial nuclear transplantation.

More recently, these investigators (Kobel, Brun, and Fischberg, 1973) transplanted nuclei of in vitro cultured adult *Xenopus* liver cells (Rafferty, 1969). The culture had undergone a minumum of 60 subcultures and was predominantly aneuploid. Despite the aneuploidy, serial transfer of the cultured *adult* nuclei led to the development of two clones that reached advanced tailbud stages and lived for about a day and a half (Stages 30 and 32, Nieuwkoop and Faber, 1967).

What conclusions may be drawn from the aneuploid cell line cloning experiment? One is that obtaining an advanced embryo is remarkable because aneuploidy is not compatible with normal development. Thus, it would seem that their methodology has extracted a near maximum of information from the in vitro cultured *adult* cell line. Further, it should be noted that they transferred not an adult liver cell nucleus but a nucleus from cultured adult liver cells, which are mitotically active. Finally, the best development was obtained by serial transplantation. These methods are certainly in harmony with those cloning procedures discussed above for *Rana* and *Xenopus*, which led to maximum embryonic development after transplantation of adult nuclei.

Other studies of maturing nuclei of *Xenopus* have been attempted but with less success. Nuclei of differentiated limbs of premetamorphic larvae were forced to become mitotically active and morphologically dedifferentiated by formation of a regeneration blastema after amputation. No larvae were produced by transplantation of the *Xenopus* blastema nuclei (Burgess, 1967, 1974; Chapter 5).

A Final Thought

Little is known concerning what permits a genome from a differentiating cell to be reprogrammed such that it mimics a zygote nucleus. Much of that which has been written about the subject relates to exceptional nuclei. The vast majority of adult nuclei *do not* promote embryonic development in cloning experiments even in the most promising systems. We infer biological mechanisms from common observations in most aspects of cell biology. Exceptional observations in amphibian cloning are the data base on which we draw conclusions about the functioning of the genetic material. Accordingly, extreme care should be used in drawing conclusions concerning observation that have intrinsic technical pitfalls (such as the propensity for gynogenesis, imperfect enucleation, etc.).

CHAPTER 7

An Omnibus Chapter: A Discussion of Nuclear Transfer Studies Not Relating Directly to Nuclear Differentiation

The nuclear transplantation technique provides an opportunity to effect combinations of nucleus and cytoplasm that will provide answers to biological questions other than the problem that generated the procedure. Some of the studies to which the technique has been addressed will be considered below.

Replicate Anurans and Urodeles with a Consideration of Histocompatibility Factors

Groups of frogs that are genetically identical ensue from nuclear transplantations of a common blastula donor. Such a group from a common nuclear donor are as similar as identical twins and are thus said to be isogenic. The terms "isogenic groups," "syngenetic series," and nuclear transfer "clone" are used synonymously in this chapter. Thus, "clone" is used here to mean a group, in lieu of its meaning as a mode of asexual reproduction (see definition of clone on pages vii-viii).

Lamentably, large groups of spontaneously occurring genetically identical vertebrates are rare. Twinning and other forms of multiple genetically identical births are not common. Inbreeding can effect a large degree of homozygosity but inbred progeny tend to be characterized by a loss of vitality. Nuclear transfer affords a means of producing a potentially unlimited number of genetic replicate animals that have none of the problems associated with inbreeding (Gurdon, 1962d). Isogenic groups of urodeles produced by cloning, as well as groups of frogs, have been reported (Gallien, Picheral, and Lacroix, 1963a).

Two isogenic groups of *Xenopus laevis* were described by Gurdon (1962d). The frogs in each group were remarkably similar to each other with the exception of two frogs in one group that were retarded in growth and sexual development. Control groups were comprised of both sexes. One sex only was found among the normal members of a syngenetic series. Color patterns of skin varied among genetically different *Xenopus*. Isogenic frogs were reported to be very similar in skin pattern within each group. However, it might be noted that pigmentation differences were not particularly obvious in the photographs of *Xenopus* (Gurdon, 1962d, figures 2 and 3). Variations in size within groups were reported but it seems reasonable to believe that environmental fluctuations in culture conditions would affect growth rates of genetically identical individuals and thus size differences would not be unexpected.

Wild-type (spotted) and Kandiyohi (mottled) skin spotting patterns in syngenetic frogs were described in McKinnell (1962b). (Pigment pattern mutants of *R. pipiens* are described in Appendix I.) The results graphically demonstrate that genetic factors are more important than environmental factors in most of the characters studied. Thus, there was total conformity in pigment pattern type within a syngenetic group, that is, only wild-type or Kandiyohi frogs were obtained within any individual group. Five clones were presented and four of the clones were comprised entirely of wild-type (spotted) frogs. A fifth clone was composed exclusively of Kandiyohi frogs. Sex was similarly determined by genetic factors. Each clone was of one sex with a single exception. A unique male (accompanied by seven female frogs) was found in one isogenic group. It was postulated that elevated laboratory temperature might have had a masculinizing effect on the development of the exceptional frog (Witschi, 1929). The size, number, and arrangement of spots on the leopard frogs varied within a clone. Despite the variation within a clone, an analysis of variance test of group differences of isogenic groups revealed that the spot number varied significantly (i.e., differences between clones are greater than differences among individuals within a clone). As in Gurdon's 1962d study, there were differences in growth rate as measured by the time required to reach the stage when the anterior limbs appear (McKinnell, 1962b).

Small isogenic groups of Burnsi (spotless) frogs (Figure 7-1) were studied by Simpson and McKinnell (1964). Striking similarities in skin pigment pattern were observed (Figure 7-2). We concluded that "environmental fluctuations, which may have included variations in recipient egg cytoplasm quality and other subtle variations not easily controlled, were not sufficient to alter the intrinsic control of the pigment patterns of these isogenic groups of frogs" (Simpson and McKinnell, 1964).

In contrast with the experiments described above that suggest a paramount role of genetics in the control of pigmentation patterns, Casler (1967) reported no significant difference in variability with regard to spot number, average spot size, and spot location and shape between clones and controls. Her frog replicate groups were impressively large. Additional experiments will perhaps resolve the differences between Casler's results and those of Gurdon and of my laboratory.

Substantial phenotypic variation in frog serum proteins, as detected by polyacrylamide gel electrophoresis was reported by Pogany (1970). The variations were comprised of numbers of proteins, quantity of protein (as revealed by staining intensity), and speed of migration of proteins in the electrophoretic gel. To what extent are these variations due to genetic diversity and to what extent are they due to the physiological state of the animal as influenced by the environment? We sought an answer to this question by examining sera of isogenic groups of frogs produced by nuclear transplantation (McKinnell, Steven, and Ellgaard, 1973; see also Wabl and DuPasquier, 1976). Mobility and intensity of staining of protein bands varied within an isogenic group but the number of bands remained constant. The significance of variation in genetic replicates is unknown but it is probably due to minute differences in feeding behavior, quality of recipient egg cytoplasm, and temperature and light regimen. Further, the method of protein separation may result in minor fluctuations in protein migratory behavior. However, we were not alone in reporting variation between

Figure 7-1. Nuclear transplant frog of mutant phenotype produced by transplanting a Burnsi diploid nucleus into an activated and enucleated ovum obtained from
a wild-type *Rana pipiens*. (From Simpson and McKinnell, 1964.)

members of a syngenetic group. Aimar and Chalumeau-LeFoulgoc
(1969) recorded slight variations in the level of serum proteins
among isogenic *Pleurodeles waltlii*. We are currently studying liver,
brain, muscle, and serum proteins of unrelated, sibling, and isogenic frogs in an effort to further characterize similarities and differences in nuclear transfer frogs.

Surprisingly, there is little direct evidence supporting the view
that all somatic nuclei of an individual blastula, older embryo, or
adult contain the same genes as were present in the zygote nucleus
(Gurdon, 1963). The migration of proteins derived from isogenic
frogs, in an electrophoretic field, provides ambiguous data for the
time being. However, the experiments with Burnsi and Kandiyohi

Figure 7-2. Diagram illustrating method for production of isogenic groups of mutant frogs. Burnsi adults (*line 1*) yield eggs and sperm from which are produced donor blastulae (*line 2*). Donor blastulae are dissociated (*line 3*) and transplanted (*line 4*) to activated and enucleated recipient ova. Adult nuclear transplant progeny (*line 5*) are obtained. The frogs, forming small groups within brackets, are syngenetic. The individuals of group A resemble each other in having immaculate dorsal body surfaces, the frogs in group B have small spots located posteriorly on their dorsal body surfaces, the pair of group C have substantial dorsal pigmentation, and the animals of group D, like those of group A, resemble each other in having immaculate backs. (From Simpson and McKinnell, 1964.)

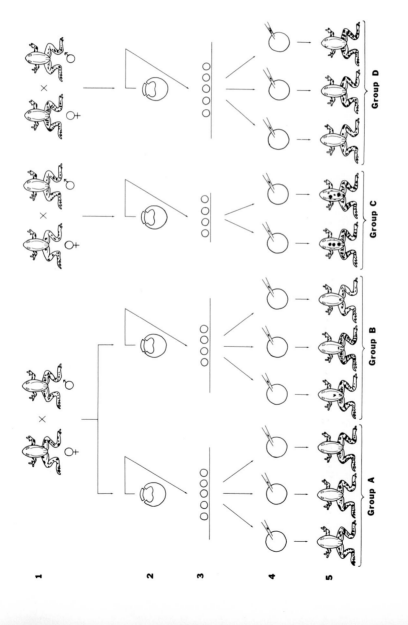

Group A   Group B   Group C   Group D

1
2
3
4
5

isogenic frogs, as well as the studies of histocompatibility to be discussed below, provide compelling evidence that, at least with respect to the two mutant genes and an unknown number of histocompatibility genes, all nuclei from a common blastula are genetically identical.

Embryonic *Xenopus* tissue implanted to immunologically competent hosts elicits a rejection reaction as in other vertebrates (Simnett, 1966). A graft of tissue from one *X. laevis* to another *X. laevis* is referred to as a homograft but a graft from one part of an individual to another anatomical location on the same individual is known as an autograft. It is significant that reciprocal grafts between members of the same isogenic group behave as autografts but reciprocal grafts between animals of separate clones are rejected as homografts (Simnett, 1964a). The grafting experiments with *Xenopus* were confirmed with experiments with nuclear transplant leopard frogs (Volpe and McKinnell, 1966).

In the leopard frog, as in *Xenopus*, an embryonic homograft elicits an immune response (Volpe and Gebhardt, 1965; see also DuPasquier, 1976). However, the rejection response is dose dependent. A large homograft will be tolerated (Gebhardt and Volpe, 1973). Neural crest grafts (Figure 7-3) exchanged between embryos of an isogenic group persist beyond metamorphosis (Figure 7-4). Similarly, grafts of dorsal skin between juvenile frogs of the same isogenic group were not rejected (Figure 7-5). Interclonal grafts, i.e., skin transplants between members of different isogenic groups were invariably rejected (Volpe and McKinnell, 1966).

Of interest is the persistence of skin grafts between diploid and tetraploid juvenile frogs of the same clone (Figure 7-6). Failure of rejection by a diploid host of a tetraploid graft lends substantial credence to the view that tetraploidy arises in nuclear transfer studies by the fusion of descendants of an implanted diploid nucleus after one mitotic division. Fusion of two diploid nuclei derived from a single implanted nucleus would result in a tetraploid frog isogenic to its diploid clonal compatriots. Tetraploidy arising by fusion of an inserted embryonic nucleus (2n) with a retained second polar body nucleus (1n) and the mature maternal pronucleus (1n), or by fusion of the inserted diploid nucleus with

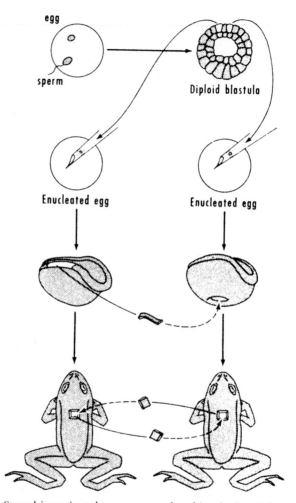

egg

sperm

Diploid blastula

Enucleated egg

Enucleated egg

Figure 7-3. Several isogenic embryos were produced by cloning. Only one pair of identical embryos is shown. When this pair reached the neurula stage of development, one lateral neural fold of one embryo was transplanted to the abdominal region of the other embryo. This was followed, when both embryos metamorphosed into frogs, by a reciprocal transplantation of pieces of dorsal skin.
(From Volpe and McKinnell, 1966.)

Figure 7-4. Three examples of host frogs bearing viable neural crest grafts in the abdominal region. The pigment cell derivatives of the transplanted neural crest express a spotting pattern typical of the dorsal skin of the leopard frog.
(From Volpe and McKinnell, 1966.)

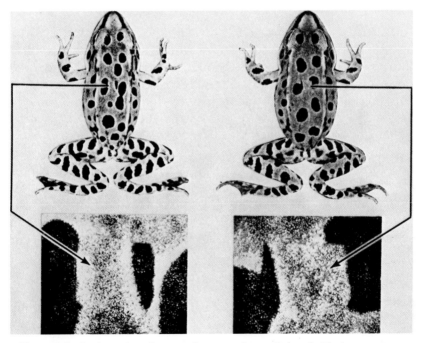

Figure 7-5. Reciprocal exchanges of square pieces of dorsal skin between two young leopard frogs of the same isogenic group. The enlarged views of the grafts reveal the viable condition of each skin graft. (From Volpe and McKinnell, 1966.)

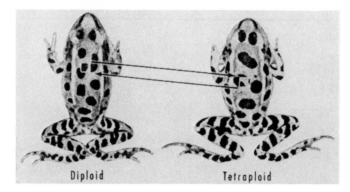

Figure 7-6. Successful exchange of skin grafts (outlined in white for clarity) between a diploid and a tetraploid leopard frog of the same syngenetic group. (From Volpe and McKinnell, 1966.)

a maternal pronucleus that has divided once followed by reconstitution to a single diploid nucleus would in both cases result in embryos that would not be homozygous with diploid nuclear transplant frogs. The study in ploidy suggests that qualitative genetic diversity, not quantitative genetic diversity, is responsible for the rejection reaction (Volpe and McKinnell, 1966).

The importance of genetic diversity to immune rejection has been probed further by Roux and Volpe (1974). They investigated the addition of genetic diversity by including an unrelated haploid set of chromosomes in a nuclear transplant series. As a consequence, some of the clone members were triploid and some were diploid. Reciprocal skin grafts between triploid and diploid clone members revealed that a single dose of histocompatibility genes was sufficient to elicit an immune response.

Current studies in my laboratory by my student Lyle Steven, Jr., are designed to utilize the nuclear transplantation procedure as a means of resolving whether sexual dimorphic histocompatibility in amphibia (Pizzarello and Wolsky, 1960) is dependent upon the presence of male antigens or male hormone.

Isogenic *Pleurodeles waltlii* with homologous (*P. waltlii*) or heterologous (*P. poireti*) cytoplasm tolerate skin grafts. Thus, histocompatibility factors are nuclear and foreign cytoplasm does not alter the immunology of skin graft tolerance (Aimar and Gallien,

1972). More recently, these authors have reversed the phenotypic sex of male nuclear transfer *Pleurodeles* by utilization of estradiol. The reversed sex females, termed "neo-females" by Gallien and Aimar, were reared to sexual maturity and mated with their isogenic but untreated male counterparts. Only male offspring developed from the crossing of the genetic replicates (Gallien and Aimar, 1974). The reversed sex study is described here because the experimental system can be utilized in an analysis of why isogenic male tissue is rejected by genetically identical (except for sex) females. Is rejection due to the presence of a Y chromosome with unique antigens or is it due to the presence of male hormone. As indicated above, we are studying the question in my laboratory.

Nucleocytoplasmic Chimeras: The Construction of Cells with Nuclei and Cytoplasm of Disparate Origin

From a manipulatory point of view, it is as simple to effect a taxonomically novel combination of nucleus and cytoplasm as it is to reconstitute a developmental system using parts from a single species. Consequently, it is not surprising that unusual unions of nucleus and cytoplasm have been realized. Indeed, the first nuclear exchange experiments were between frogs and toads; unfortunately, the transplanted eggs died (Rauber, 1886). Viable nucleocytoplasmic hybrids have been effected from mutants, subspecies, and species. Certain of the operated eggs develop to adult nucleocytoplasmic chimeras, others have less potential for growth.

Nuclei of the Japanese newt, *Triturus pyrrhogaster*, were inserted into enucleated eggs of the northern leopard frog, *Rana pipiens*. It is perhaps not surprising that neither a jumping newt nor a tailed frog ensued from this Nippon-Yankee combination—only partial blastulae were formed. However, some cells of the hybrid blastulae contained chromatin organized into "nuclei." The chromatin was increased relative to the amount introduced by the donor nucleus showing that DNA replication had taken place (Briggs and King, 1955). Intergeneric transfers, *R. temporaria* nuclei to *X. laevis* cytoplasm, also failed to yield viable embryos (Fischberg, Gurdon, and Elsdale, 1958b).

Pigment pattern mutant genes of *R. pipiens* are entirely compatible with cytoplasm derived from wild-type leopard frogs. Nuclei from the spotless mutant (Burnsi) and the mottled mutant (Kandiyohi) (Appendix I) replicate normally in enucleated ova obtained from spotted (wild-type) leopard frogs (McKinnell, 1960; Simpson and McKinnell, 1964). A Kandiyohi nucleocytoplasmic hybrid frog was reared to sexual maturity and backcrossed to a homozygous recessive leopard frog. Wild-type and mutant Kandiyohi phenotypes were recovered in the backcross indicating that the somatic nucleus initially transplanted was heterozygous with respect to the mutant Kandiyohi gene (McKinnell, 1962b). The results of the backcross indicate the feasibility of genetic analysis of somatic vertebrate cells.

Kandiyohi is dominant to twice the usual number of wild-type alleles. The Kandiyohi phenotype is expressed in triploid frogs produced by transplanting a heterozygous (k/+) nucleus to an activated (but *not* enucleated) recipient egg (+). The resulting triploid frog (k/+/+) has the characteristic mottling of the Kandiyohi mutant (Figure 7-7) (McKinnell, 1964). These experiments with Burnsi and Kandiyohi indicate that the pigment pattern variants are under nuclear, not cytoplasmic, control. Accordingly, they were suggested as suitable for nuclear markers for transplantation experiments (Chapter 2). In the same fashion, a dominant black gene was used as a nuclear marker in transplantation experiments of axolotl eggs that were obtained from homozygous recessive (white) females. Expression of the black gene was evidence that the introduced nucleus was guiding development of the transplant embryo (Signoret, Briggs, and Humphrey, 1962; see also Dasgupta, 1969, for a similar use of a mutant gene as a nuclear marker).

"Subspecific" nucleocytoplasmic chimeras were first reported by Sambuichi (1957a). One hybrid tadpole neared metamorphosis from a total of 262 operated eggs. The tadpole displayed morphological characteristics of the nuclear-donor subspecies. Subsequently, nuclear transfer experiments were reported in the reverse direction; i.e., nuclei of *Rana nigromaculata nigromaculata* were transplanted into enucleated eggs of *R. n. brevipoda* (Sambuichi, 1961). Two metamorphosed frogs ensued from the later experiment. It should be noted here that the "subspecific" nucleocytoplasmic hybrids reported by Sambuichi were probably interspecific nuclear

Figure 7-7. Triploid Kandiyohi frog produced by inserting a diploid somatic nucleus into an activated (but *not* enucleated) host ovum. (From McKinnell, 1964.)

transfers. *R. nigromaculata brevipoda* was designated a valid species, *R. brevipoda*, by Kawamura (1962) (see also Moore, 1960b, p. 11).

The earlier studies of Sambuichi were continued by Kawamura and Nishioka. Fourteen diploid tadpoles produced from enucleated *R. nigromaculata* eggs and diploid blastula *R. brevipoda* nuclei completed metamorphosis. Thirteen of these frogs resembled the nuclear donor in most respects except that dorsal body spots were intermediate between the nuclear and cytoplasmic species and the hybrid frogs were delayed at metamorphosis. The single exception among the 14 frogs was judged to have developed by parthenogenesis (Kawamura and Nishioka, 1963a). Five tadpoles developed from the reciprocal nucleocytoplasmic cross (Kawamura and Nishioka, 1963a).

An amphidiploid (a tetraploid animal composed of diploid sets of chromosomes from both species) frog produced by inserting a diploid blastula nucleus of *R. brevipoda* into a fertilized *R. nigromaculata* egg was reported by Kawamura and Nishioka (1963b). The frog was male and was reproductively normal. Triploid offspring were produced when it was mated with diploid *R. nigromaculata* or *R. brevipoda* females.

R. nigromaculata and *R. brevipoda* are closely related and have been referred to as sibling species. Nucleocytoplasmic hybridization of these two species indicates a high order of compatibility with most characters of the resulting offspring being of the nuclear type. Some cytoplasmic effect has been described. The effect of foreign cytoplasm on the expression of genetic characteristics should be greater if two species more remotely related to each other than the two described above are studied. Accordingly, nucleocytoplasmic hybrids were made using *R. japonica* and *R. ornativentris*. Fourteen frogs were produced in these nuclear transfer experiments involving species that differ in chromosome number (Kawamura and Nishioka, 1963c). These studies are continuing and recently, nucleocytoplasmic hybrids between *R. japonica* and *R. temporaria*, *R. esculenta* and *R. brevipoda*, and *R. brevipoda* and *R. plancyi chosenica* have been reported (Nishioka, 1972a, b, c, Kawamura and Nishioka, 1972).

Two subspecies of the South African clawed frog, *Xenopus laevis*, have been studied as nucleocytoplasmic chimeras. *X. l. laevis* as a juvenile frog has a dark green or blackish ground color with black mottling. *X. l. victorianus* is a light yellow-green with a very fine freckling of dark green to black. Heterotypic frogs derived from *victorianus* nuclei and *laevis* cytoplasm were reported to be identical to control *victorianus*. Gastrula nuclei of *victorianus* that were serially transplanted four times into eggs of *laevis* gave rise to entirely normal frogs resembling *victorianus*. The serial transfers involved at least 60 cell cycle replications of *victorianus* chromosomes in non-homologous cytoplasm with no detected effect on the *victorianus* phenotype of the heterotypic frog. Replication of *victorianus* chromosomes in foreign cytoplasm for 30 nuclear divisions with retransfer back to homologous cytoplasm also did not affect normal development. Serial transplantation in heterologous cytoplasm did not affect reproductive capacity of the nuclear transplant frogs. Nuclear control of subspecific phenotype as well as nucleocytoplasmic compatibility of two subspecies of *Xenopus* thus have been convincingly demonstrated (Gurdon, 1961). More recently, nucleocytoplasmic hybrids between *X. laevis laevis* and *X. laevis petersi* have been reported (Ortolani, Fischberg, and Slatkine, 1966).

The experiments discussed thus far record combinations of nuclei and cytoplasm that are completely or almost completely viable. One would predict less viability in the synthetic embryo as the relationship between the nuclear and cytoplasmic species decreases (see review by Subtelny, 1974).

Hybrids of bullfrog (*Rana catesbeiana*) sperm and leopard frog (*R. pipiens*) eggs develop to a late blastula or early gastrula stage and arrest (Moore, 1941). Nuclei from bullfrog and leopard frog hybrids, when implanted into enucleated leopard frog eggs, result in embryos that arrest at the same stage as the egg-sperm hybrid. This is true of embryos that ensue from hybrid donor nuclei obtained before donor arrest (16 hours), at the time of arrest (26 hours), and 10 hours after the time of arrest (36 hours). Transfer of hybrid nuclei that have been arrested in their development for 20 or more hours (44- to 48-hour hybrid donors) fail. The transfer of the 36-hour hybrid nuclei demonstrate that hybrid arrest is not due to the inability of hybrid cells to replicate, for, when transplanted to enucleated host eggs, cell division resumes (King and Briggs, 1953).

Rana pipiens and *R. palustris* are morphologically similar and are sufficiently similar genetically that their egg-sperm hybrids develop to metamorphosis (Moore, 1941). Surprisingly, however, androgenetic haploid hybrids between the species develop less well than conventional androgenetic haploids of either of the species (Moore, 1950). It would seem, therefore, that a nucleus composed of chromosomes from both species has a developmental potency greater than a nucleus composed of genetic material from one species replicating in the cytoplasm of another. This hypothesis was tested in an elegant study by Hennen (1965).

Embryos that arrested at postneurula stages ensued from the insertion of diploid *pipiens* blastula nuclei into *palustris* eggs. The developmental potency of the *pipiens* nuclei was not altered by replication in foreign cytoplasm. Although the *pipiens* nuclei replicating in *palustris* cytoplasm were subject to an arrest, these nuclei at the blastula and late gastrula stage were competent to direct normal development when returned to homologous cytoplasm. As would be expected, normal chromosomes characterized both the arrested nucleocytoplasmic hybrid and the embryos that resulted from retransplantation to *pipiens* cytoplasm (Hennen, 1965, 1967, 1974). Diploid *R. palustris* nuclei inserted into enucleated *R. pipiens* cytoplasm develop no further than a postneurula stage even though there is no detectable karyotypic change in chromosome constitution (Hennen, 1972). Early gastrula ectoderm of the hybrid *R. palustris* nuclei in *R. pipiens* cytoplasm, grafted to normal *R. palustris* hosts, heal, survive for a few days, but are not enhanced in their developmental capacity (Hennen, 1973).

Can a diploid set of chromosomes (*pipiens*) that is itself incompetent to direct normal development of heterologous cytoplasm (*palustris*) be enhanced in its developmental potency by fusion with a haploid set of chromosomes (*palustris*) that is itself incompetent to guide development beyond but the earliest embryonic stages? The answer to this question that asks in developmental terms "can the halt lead the blind?" is affirmative. Diploid nuclei from *pipiens* were transferred to *non*-enucleated eggs of *palustris* — triploid nucleocytoplasmic hybrids ensued from this combination that developed through metamorphosis (Hennen, 1964). This experiment of Hennen is similar in experimental design to those of Subtelny (discussed in Chapter 3) wherein homozygous diploid nuclei, fated to an early embryological death, are fused to a haploid set of chromosomes, which when replicating alone are destined to an early arrest, resulting in a viable triploid genome that develops through metamorphosis. Thus, here are two instances of genomes which when operating in consortium accomplish more than either are competent to alone. These studies are relevant to other experiments of Subtelny (see Chapter 3) wherein endoderm nuclei are *not* enhanced in their capacity to promote development after fusion with maternal genetic material. Whatever is the nature of the limitations in developmental capacity imposed on endoderm nuclei, it is not similar to mechanisms that lead to arrest in the nucleocytoplasmic hybrids of Hennen and the homozygous diploids of Subtelny.

Nuclear transplantation between two species of *Xenopus* was reported by Gurdon (1962c). *Xenopus tropicalis* ranges in the Congo basin and Northern Angola and has eggs about one-fourth the volume of the more widely distributed *Xenopus laevis*. The frogs hybridize neither in nature nor in the laboratory. Unlike subspecific nucleocytoplasmic hybrids in *Xenopus*, *X. tropicalis* nuclei with *X. laevis* cytoplasm blocked at a neurula stage and the reverse transfer (*X. laevis* nuclei and *X. tropicalis* cytoplasm) arrested at the late blastula or early gastrula stage. *X. laevis* nuclei that replicated in *X. tropicalis* cytoplasm and were then serially retransplanted back to homologous cytoplasm continued to arrest at an early developmental stage. The effect of foreign cytoplasm on a transplanted nucleus is a change in the developmental capacity of the nucleus. The change is heritable through mitosis, is species specific (i.e., nuclei of different species are differentially affected by residence in heterologous cytoplasm), is complete at the early blastula stage, and is not reversed after serial transfer in its homologous cytoplasm for six transplant generations (ca. 70 nuclear divisions) (Gurdon, 1962c).

Sládeček and Mazáková-Štefanová (1965) reported transplanting nuclei from the axolotl (*Ambystoma mexicanum*) into nucleated *Triturus vulgaris* eggs. Three triploid larvae developed. The authors doubted that the larvae were the result of nuclear proliferation of the host nuclei alone but suggested that an immunological analysis might provide a definitive answer.

Success in transplanting a living nucleus in nuclear transfer studies has been judged by appropriate ploidy of the resulting embryo, or in some cases, pigment patterns indicating the participation of the implanted nucleus. Immunological techniques have been used only once for the detection of species-specific antigens characteristic of the nuclear donor as evidence of success in transplantation.

A diploid embryo with *Triturus alpestris* antigenicity (as detected with Ouchterlony diffusion plates), produced with the nucleus of *T. alpestris* implanted into enucleated *T. vulgaris* cytoplasm, was considered to have developed after successful nuclear transfer. A very low yield of developing embryos attested to the extreme difficulty in utilizing these species for nuclear transplantation (Sládeček and Romanovský, 1967).

What can one conclude from the results of the nucleocytoplasmic chimera experiments? Nucleocytoplasmic compatibility seems highest when the taxonomic distance between donor nucleus species and recipient cytoplasmic species is small. Thus, normal transplant animals are produced as nucleocytoplasmic hybrids between mutants, subspecies, and some closely related ("sibling") species within a genus. Transplanted nuclei arrest when the chimeras are formed between most species and in all intergeneric combinations.

Interspecific nuclear transfer described above was designed, for the most part, to reveal to what extent the genome of one species can replicate normally in the cytoplasm of another species. The end point of the experiments were embryos or adults, which were a measure of nucleocytoplasmic compatibility. There is emerging another kind of nucleocytoplasmic hybrid, not designed to give rise to an embryonic system, but rather designed to provide insight into the nature of control mechanisms regulating gene activity. Multiple HeLa cell nuclei have been inserted into oocytes of *Xenopus*. The mammalian nuclei swell to many times their original size; they synthesize RNA; and perhaps what is more striking, at least some of the HeLa informational RNA is translated into HeLa protein. The utilization of *Xenopus* oocytes as living test tubes is described in Gurdon, DeRobertis, and Partington (1976).

Which Is More Sensitive to Ionizing Radiation:
Nucleus or Cytoplasm?

A study of moderately irradiated eggs and blastulae (γ-rays of a ^{60}Cobalt source resulting in a 783 R dose per egg or blastula) revealed that damage to the nucleus is the primary reason for the poor development of irradiated amphibians (*Rana nigromaculata*). Three kinds of experiments were performed: (1) transplantation of a normal blastula nucleus to a normal enucleated egg, (2) transplantation of a nucleus obtained from an irradiated blastula into a normal enucleated egg, and (3) transplantation of a normal blastula nucleus into a enucleated egg that had been irradiated. Transplant frogs that could develop to metamorphosis developed from group 1 (the controls) and group 3. No embryos developed beyond the muscle response stage in group 2 (Sambuichi, 1964). Similar conclusions were drawn in a study of heavily irradiated eggs by the same author. Cleavage was impeded by heavy irradiation of the cytoplasm but normal tadpoles were obtained in the control group with unirradiated nucleus and cytoplasm, and in transplanted eggs that had received 49,200 R and 73,800 R. Gastrulation did not occur in transplant eggs that had received 98,400 R. Frogs were reared to metamorphosis only from the control group and from the transplanted embryos whose cytoplasm had received 49,200 R. It is clear, therefore, that the cytoplasm can withstand enormous (49,200 R) doses of γ-irradiation such that an implanted normal nucleus can result in a metamorphosed frog. Further, a dose twice that size (98,400 R) does not prevent cleavage by a normal implanted nucleus. However, the damaged cytoplasm in the latter case is no longer able to support a normal nucleus. The nature of the damage to cytoplasm from the enormous γ-irradiation is not known (Sambuichi, 1966).

The Effect of a Radiomimetic Substance on the Nucleus and Cytoplasm of Amphibian Embryos

Nitrogen mustards are mutagenic and they interfere with amphibian embryonic development (Bodenstein, 1948). It would seem, therefore, that these substances would interfere with nucleic acid metabolism during early development and, further, that the disruptions to the embryo could be analyzed by the nuclear insertion procedure.

Eggs that served as recipients for normal nuclei and blastulae that served as nuclear donors to normal eggs were treated with nitrogen mustard. Two kinds of nuclear damage were described. A direct effect caused by exposure to the agent and an indirect effect on a normal nucleus that replicated in treated egg cytoplasm were reported (Grant, 1961). Direct and indirect damage was found to occur in haploid as well as diploid nuclei. Abnormalities in number and form were found in chromosomes associated with abnormal development in most cases. However, some experimental embryos were abnormal but revealed no detected chromosomal alterations. It was suggested that in the latter situation, submicroscopic chromosomal lesions resulted from the treatment (Grant and Stott, 1962).

Surprisingly, the two reports of Grant comprise all of the amphibian nuclear transplantation literature on the effect of drugs and other agents that affect nucleic acid metabolism.

Polyploidy Produced by Nuclear Grafting

Amphibians have a physiology that is not incompatible with polyploidy. They are perhaps best suited among vertebrates in providing experimental material for the study of the effect of increased sets of chromosomes upon development. Nuclear transplantation is a convenient method of producing polyploid animals and has been exploited for this purpose.

Diploid nuclei transplanted to non-enucleated eggs of *R. japonica* yielded triploid embryos and diploid nuclei transplanted to fertilized eggs of the same species yielded tetraploid embryos (Sambuichi, 1959).

Tetraploid *Xenopus laevis* were produced by transplanting diploid nuclei. The introduced nucleus divided and fused with itself without the egg cleaving resulting in a tetraploid nucleus. Evidence for lack of the egg nucleus participating in development was demonstrated because the introduced nucleus had only one nucleolus (instead of the normal two). The tetraploid nucleus had two nucleoli attesting to nuclear fusion of mitotic daughters of the introduced nucleus. The tetraploid frogs were normal in morphology (Gurdon, 1959).

Triploid frogs with a mutant pigment pattern were produced by nuclear transplantation. The experiment was designed to test if the dominant gene Kandiyohi is expressed in the presence of two wild-type alleles. Homozygous recessive nuclei were transferred to unfertilized (nucleated) eggs of Kandiyohi females. Blastulae nuclei that were either heterozygous for Kandiyohi or homozygous recessive were implanted to eggs of wild-type females. Both Kandiyohi as well as wild-type were among the triploid frogs produced. It is thus clear that the mutant gene can indeed express itself in the presence of an additional recessive allele. Further, the experiments provide genetic evidence that a diploid somatic nucleus (the transplanted blastula nucleus) can fuse with a gamete nucleus (the maternal pronucleus). Nuclear fusion in egg cytoplasm is ordinarily limited to the union of gamete nuclei. The experiment of a heterozygous diploid Kandiyohi nucleus transplanted into the egg of a recessive female with a triploid Kandiyohi frog ensuing indicates that triploidy cannot possibly have occurred by gynogenesis (fusion of the mature maternal gamete nucleus, 1n, with a repressed first polar body, 2n; or a divided maternal gamete nucleus, 1n + 1n, joined with a repressed second polar body, 1n; which in either case would result in triploidy) because the only source of the Kandiyohi gene was the transplanted nucleus. There is no possible way the introduced nucleus could have become triploid other than by fusion with the maternal nucleus (McKinnell, 1964). The genetic proof of nuclear fusion complements the elegant cytological studies showing nuclear fusion by Subtelny and Bradt (1963).

More recently, triploid nuclear transplant tadpoles have been produced by transplanting nuclei of triploid tumor cells (see Chapter 4). The experimental design of the triploid tumor study utilizes polyploidy as a nuclear marker. Triploid cells of triton *Pleurodeles waltlii* when transplanted to activated and enucleated eggs of the same species result in larvae that are predominantly triploid. A few tetraploid larvae were detected. The tetraploid larvae were interpreted as a consequence of the failure of ultraviolet enucleation procedure. The authors suggest that triploidy would be a suitable marker in nuclear transplantation experiments (Gallien, Picheral, and Lacroix, 1963a). Their suggestion was heeded in the tumor transfer experiments of six years later.

APPENDIXES

"His instruments are simple . . ."

Preface to the Appendixes

"His instruments are simple: glass rods drawn to a point, glass tubes which can be used as fine pipettes, or loops of children's hair. His experimental material consisted of the eggs of newts and frogs."

The quotation cited above would be an appropriate description of a contemporary nuclear transplanter's instruments and material. The words were uttered not in the seventies, however, but in the mid-1930s. They are to be found recorded in the presentation speech awarding the Nobel Prize in Physiology or Medicine to *Herrn Geheimrat* Professor Dr. Hans Spemann (Figure A-1) in 1935 (Häggquist, 1965).

There is a philosophical thread that may be followed from the amphibian embryo experiments of Spemann to present-day cloning. As the experiments are conceptually similar, it should not be surprising to find that so also are the contrivances and implements necessary for the operations. Microsurgeons have always had to possess skills not only in the wielding of micropipettes and microneedles but also in the manufacture of them. There is no commercial source of fine glass tools used in experimental embryology. Spemann (1923) described how he made his simple instruments. How they are made today is described in Appendix II.

Although modern tools are about the same as the tools in the days of Spemann, tool making has become more sophisticated.

Figure A-1. *Herrn Geheimrat* Professor Dr. Hans Spemann. Photograph taken in 1914 at about 45 years of age. (From: Friedrich Wilhelm Spemann, *Hans Spemann: Forschung und Leben.* Stuttgart: J. Engelhorns, 1943.)

The appendixes are concerned with the refinements in fashioning implements needed for the delicate procedure of nuclear transfer, with the selection and husbandry of experimental animals, and with other specifics, and, in addition, tell how to perform a nuclear transplantation operation. This section is intended as a brief vade mecum for one kind of amphibian microsurgery, namely, the kind of microsurgery that involves the thrusting of a living nucleus into egg cytoplasm bereft of chromosomes in such a manner that a viable nucleocytoplasmic combination ensues.

While it would be inappropriate to dedicate a vade mecum to

a group of cold-blooded vertebrates, perhaps a kind word or two would not be out of order. A number of years ago, Weiss (1939) had some nice things to say about amphibians which included:

Although the echinoderm egg deserves a large share of the credit for the modern development of Analytical Embryology, it is doubtful whether even the most intense study of this single object could ever have brought us such specific knowledge about fundamentals of development as we now possess. For, the variety and specialization of organs and tissues are fairly modest in an echinoderm larva; at any event, they are inferior to those of even the simplest vertebrate. Under these circumstances it has been a veritable boon to our science that germs of vertebrates, especially amphibians, have proved to be just as accessible to experimentation as the echinoderm eggs. The experiments have also been extended to the germs of *birds*, *fishes*, and *mammals* among the vertebrates and to *insects* among the arthropods, but none of this work has as yet attained such prominence as that done on *amphibians*.

The Northern Leopard Frog, *Rana pipiens*

Two decades after the first successful nuclear transplantation experiments that utilized living cells and eggs obtained from the northern leopard frog, *Rana pipiens* (frontispiece and Figures A-2, A-3, and A-4), that species is still the anuran of choice for most American embryologists concerned with the technique.

The popularity of the leopard frog for nuclear transplantation studies derives from the low cost of commercially obtained leopard frogs, convenience in storage and maintenance of animals, simple rearing techniques of experimentally produced animals, availability of eggs and sperm from about mid-October through the following June, the relatively uncomplicated procedures needed to activate and enucleate the recipient eggs coupled with simple methods for dissociating embryonic cells, and large chromosomes that are relatively easy to study. The appendixes of this book will be concerned with the exploitation for research purposes of these useful characteristics of *R. pipiens*.

Morbidity of many frogs is the most important disadvantage for researchers. Many commercially obtained and field-collected leopard frogs from the United States and Canada have a reduced vitality. The condition is manifested by the inability of many frogs to withstand storage at low temperature for prolonged periods. More important, it is becoming increasingly difficult to obtain fe-

Figure A-2. A northern leopard frog, *Rana pipiens*, collected by the author in Minnesota. Photograph by Gordon A. F. Dunn (from Nace et al., 1974).

Figure A-3. The Burnsi mutant of the common leopard frog, *Rana pipiens*. Photograph by Gordon A. F. Dunn (from Nace et al., 1974).

Figure A-4. A Kandihoyi mutant of the common leopard frog, *Rana pipiens*. Photograph by Gordon A. F. Dunn (from Nace et al., 1974).

male frogs with eggs that fertilize well and that develop to metamorphosis. While only a decade ago, 95% or more normally fertilized control eggs developed without incident to metamorphosis, today it is not unusual for many fertilized eggs to fail to cleave at all. Further, many of the eggs that do cleave die before metamorphosis. The dearth of vigorous frogs with good eggs has elicited wide interest and some disagreement concerning proper treatment of afflicted animals (Amborski and Glorioso, 1973; Emmons, 1973; Gibbs, 1973; Papermaster and Gralla, 1973). While the scarcity of sufficiently salubrious frogs is a major problem to researchers, it need not limit experiments if attention is paid to the well being of research animals and their ailments. Proper husbandry to assure good results will be discussed in this chapter.

Lest we delude ourselves into thinking that abundance in leopard frogs means that they are an inexhaustible resource, it may be well to consider that one leopard frog form, *R. pipiens fisheri* (designated *R. fisheri* on map, Figure A-5), the Vegas Valley Leopard

Frog, is described as endangered (i.e., in immediate danger of extinction); it is "perhaps the only amphibian to have become extinct in historic times in the United States" (*Red Data Book*, Vol. 3, 1970, International Union for the Conservation of Nature and Natural Resources, Survival Service Commission). No Vegas Valley Leopard Frog has been known to be collected since 1942. The reason for the decline or loss of this leopard frog is thought to be changes in the frogs' habitat due to environmental water control.

What Is Rana pipiens? *Where Does It Come From? Some Comments on Commercial Frogs and Remarks on Why One Must Know the Origin of Experimental Animals*

Many experiments were described in the first section of this book that utilized eggs and embryos of the northern leopard frog, *R. pipiens*. The leopard frog referred to was, for the most part, the leopard frog of commerce. The "frog of commerce" is not a subspecies or race of *Rana pipiens* but consists of animals that become the wares of businessmen because of the abundance of the frog, accessibility of it to commercial collectors, and a ready market created historically by the needs of gourmets who took gustatory delight in feasting on the posterior appendages of the beasts and, more recently, by the needs of teachers and researchers. The low cost of these beautiful commercially available anurans has the questionable virtue of keeping many experimental biologists in the laboratory and out of the field.

Personal communication with frog merchants located in northern Vermont has assured me that until relatively recently Vermont vendors retailed leopard frogs caught in the Lake Champlain area of Vermont, New York, and Quebec Province. Wisconsin merchants sold frogs collected primarily in Wisconsin, neighboring Minnesota, and eastern South Dakota.

Now, however, traditional sources of frogs are being depleted rapidly. Bulldozers fill breeding ponds with dirt and marshy areas are drained as agricultural reclamation and urban development encroaches upon previously undisturbed areas. Highway construction takes its toll of anuran populations indirectly by making rural undeveloped areas more accessible to city people (which encourages "development" of lake shore sites for recreational purposes) and directly by the draining of land in the construction of roadbeds and the interruption of frog migration routes. There is an enormous slaughter of frogs on the roadways after early evening rains in the spring and autumn in Minnesota and other northern states. One may surmise, with little chance of contradiction, that loss of living space will decrease frog populations. One well-established and long-time Minnesota frog dealer advised me that he believes that chemical fertilizers and agricultural pesticides are of at least equal or perhaps greater importance as a cause of dwindling frog populations as is decreased living area for the species. He related to me witnessing masses of dead frogs after rain along the margins of ponds adjacent to treated fields. Presumably the run-off water carried sufficient toxic chemicals to cause the death of the frogs. Whether or not the frog dealer's observations will withstand more careful scrutiny with respect to the actual cause of the frog kill, there is no question that frogs are found in reduced numbers in many northern states. For these reasons, commercial frog collectors are having to abandon collecting sites of long standing and seek the vanishing batrachian in ever more remote localities.

Wisconsin dealers now sell frogs collected in Mexico (Bagnara and Frost, 1977) and Canada, as well as North and South Dakota and Minnesota. Similarly, a Vermont vendor told me that he markets frogs obtained from a Wisconsin dealer when he (the Vermont dealer) has exhausted his stock of local frogs.

While not knowing the geographic origin of a leopard frog is not likely to be catastrophic for a high school biology dissection, it may be ruinous for a frog cloning experiment. Why is this so?

Systematists have varying views as to what the leopard frog is. The frog has been thought of either as one widely distributed, highly variable species with varying incompatibility between populations associated with geographic separation, or as a group of sibling species.

The fact that leopard frogs are not all alike was recognized over 80 years ago. Cope (1889) distinguished four subspecies of *R. pipiens*. Kauffeld (1937) suggested that leopard frogs comprised three species, the New England and Canadian *R. brachycephala*, the southern *R. sphenocephala*, and, distributed between the other two species, *R. pipiens*. It is instructive to refer to the map (Figure A-5) of Wright and Wright (1949), which illustrates their concept of the distribution of various *Rana* forms with particular reference to the leopard frog. The map shows distribution of four subspecies of the leopard frog (viz., *R. p. pipiens*, *R. p. berlandieri*, *R. p. brachycephala*, and *R. p. sphenocephala*). The authors assert that they have "suspended judgement" concerning whether or not *R. pipiens* is one species or many subspecies. "*R. p. burnsi*" and "*R. p. kandiyohi*" are incorrectly designated on the map as subspecies. They are pigment pattern mutants of the northern leopard frog (see following section on genetic variants of the leopard frog). Twenty aspects of morphology were evaluated in frogs obtained from a number of localities in a study that led Moore (1944) to the conclusion that leopard frogs comprise a single species. In this respect, Moore's view was in harmony with Dickerson (1906) who, concerning *R. virescens* and *R. v. brachycephala*, wrote: "It is not possible, in living material obtained from New England, New York, Michigan, Minnesota, Wisconsin, Colorado, Texas, and Arizona, to make a distinction into these subspecies. The variation of the frogs is remarkable, but no fundamental characteristic (such as proportionate length of head and body, leg measurement, etc.) remains stable when a large series of frogs from adjoining districts are examined." Dickerson, however, recognized the southern leopard frog, *R. sphenocephala*, as a separate species. Pace (1974) believes that populations of leopard frogs from the United States are composed of *numerous* species, four of which are *R. pipiens*, *R. utricularia*, *R. berlandieri*, and *R. blairi*.

One writer concluded, after reviewing the status of the *Rana pipiens* complex, that "one could come to the reasonable conclusion that these amphibians are not the best animals to use in experimental research at the present time" (Brown, 1973). While I agree with Brown that people who buy leopard frogs carelessly are subjecting their research results to a degree of vulnerability (because of the complexity of the group), I believe that the characteristics of certain *Rana pipiens* populations make them exceedingly useful to the cloner.

It is not crucial to the cloner which of the interpretations of the *R. pipiens* complex—whether it is one variable species or many

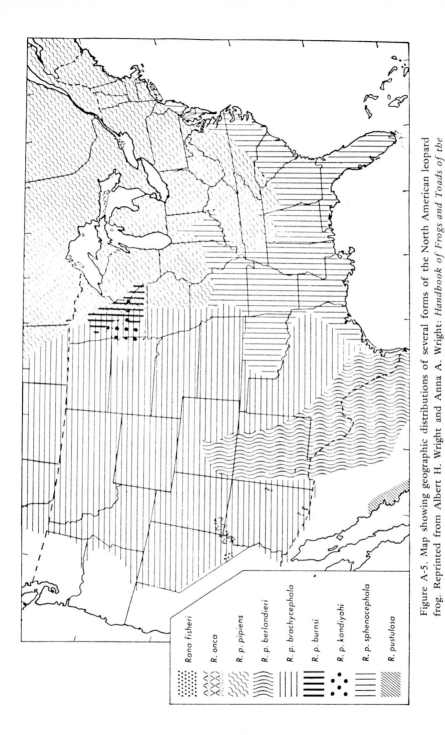

Figure A-5. Map showing geographic distributions of several forms of the North American leopard frog. Reprinted from Albert H. Wright and Anna A. Wright: *Handbook of Frogs and Toads of the United States and Canada*. Copyright © 1933, 1942, 1949 by Comstock Publishing Company, Inc. Used by permission of Cornell University Press.

Rana fisheri

R. onca

R. p. pipiens

R. p. berlandieri

R. p. brachycephala

R. p. burnsi

R. p. kandiyohi

R. p. sphenocephala

R. p. pustulosa

forms that vary in physiology and reproductive compatibility — is correct. The dissimilarities are real in either case and are particular-ly evident when the genome of one form replicates in the egg cyto-plasm of another (e.g., androgenetic haploid hybrids, Moore, 1969). A nuclear transplanter may experience unexpected lethality not directly due to the developmental processes he is studying because of the inadvertent production of inviable nucleocytoplasmic com-binations. Accordingly, it is of the utmost importance for the practitioner of nuclear exchange to know from where his frogs come.

Not only nucleocytoplasmic incompatibility may be expected but profound physiological differences are found among frog popu-lations. In the extreme these differences even affect the chronolo-gy of induced ovulation. Thus, while northern leopard frogs spawn in the spring (and eggs are obtainable by induced ovulation from October at least through the following spring), leopard frogs from northwest Sinaloa, Mexico, can be induced to ovulate primarily during July and August (Bagnara and Stackhouse, 1973). A frog experimenter could have healthy Minnesota leopard frogs in the summer and healthy Sinaloan leopard frogs in the winter and nev-er succeed in getting eggs.

One way to eliminate the problem of disputed or unknown ori-gin of frogs is for the investigator to collect them from the field himself. Communication with the dealer is desirable if collection in the field is not feasible. My experience with dealers is that they are anxious to cooperate with the scientist. Most will supply frogs collected from a restricted locality if requested and if it is possible to do so. Confirmation from the dealer of the frogs' origins is necessary because the postmark on a box of purchased frogs does not necessarily reveal the inhabitants' patria.

Genetic Variants of the Leopard Frog

Although *R. pipiens* comprise an exceedingly complex species or group of species, mutations that have been described are limited in number and primarily affect pigmentation (Browder, 1975). Some genetic variants (mutations) of the leopard frog are listed because they may be of use as genetic tags or markers in certain experiments (see discussion in Chapter 2).

Burnsi. Leopard frogs are so called because of spots on the dorsal surfaces of the body and limbs. Dickerson (1906) illustrated a spotless leopard frog captured at White Bear Lake, Minnesota, and there is a superb color illustration of the form in Merrell (1975). The frog was described as a new species, *R. burnsi*, by Weed (1922) but breeding experiments revealed that the spotless condition is due to a single dominant Mendelian gene (Moore, 1942) that results in fewer melanophores than are found in wild-type frogs (Smith-Gill, 1973). The Burnsi phenotype varies from immaculate on all dorsal surfaces to spotted limbs with one to a few spots on the back (Figures A-3 and A-6).

Although spotless leopard frogs occur occasionally in Vermont and other parts of North America, they occur at a high frequency (as much as 10%) in central Minnesota, western Wisconsin, eastern

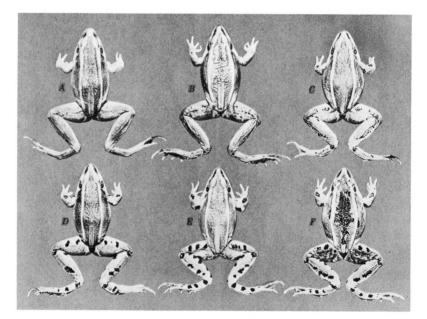

Figure A-6. Variation of spotting in Burnsi variants of the
leopard frog, *Rana pipiens*. (From Volpe, 1961.)

South Dakota, and possibly northern Iowa (Breckenridge, 1944; Merrell, 1965; McKinnell and McKinnell, 1967; McKinnell and Dapkus, 1973).

Kandiyohi. *R. pipiens* with vermiculate mottling between dorsal spots are known as Kandiyohi (Figure A-4) (see Merrell, 1975, for color illustration). Mottled frogs were described as a new species, *R. kandiyohi*, by Weed (1922) but subsequently the trait was shown to be inherited as a dominant gene by Volpe (1955).

Kandiyohi frogs have been reported only from west central Minnesota and the eastern Dakotas (Breckenridge, 1944; Merrell, 1965; McKinnell and McKinnell, 1967; McKinnell and Dapkus, 1973).

Speckle. Speckled frogs have a reduced number of iridophores, which in the heterozygote results in a dorsal pigment pattern broken by numerous olive-gray speckled markings (Figure A-7). The homozygote is similar to melanoid (described below). The speckled frog is found on the shores of Lake Manitoba, Canada, and the trait is incompletely dominant with variable expression. Leopard frogs similar to speckle but with an even greater reduction in guanophores and xanthophores and with iris pigmentation almost

Figure A-7. Heterozygous speckled leopard frog. The dorsal and lateral pigment pattern of the frog is broken by numerous olive-gray speckled markings due to the reduction in numbers of iridophores. Photograph courtesy of Dr. Leon W. Browder.

absent have a phenotype designated extreme speckle. The gene is dominant (Browder, 1968a, b).

Blue. Blue *R. pipiens* have been reported (Berns and Narayar, 1970) (for color photograph, see Browder, 1968a, figure opposite p. 163, and cover of *BioScience*, October 15, 1971, Vol. 21, No. 20). The blue condition is thought to be recessive because blue frogs crossed with the wild-type yield only wild-type progeny (Browder, 1968a).

Albino. Albino frogs of several species including *R. pipiens* have been reported (Rose, 1962; Smith-Gill, Richards, and Nace, 1972; see color illustration on cover of *BioScience*, October 15, 1971, Vol. 21, No. 20; and illustrated in Merrell, 1975). Browder crossed an albino leopard frog with a wild-type frog; the result was an all wild-type F_1 generation. A $3:1$ ratio was obtained in the F_2 generation showing that albino in the leopard frog is recessive (Browder, 1972).

Melanoid. Frogs having melanophores but deprived entirely of guanophores and xanthophores were produced among the progeny in a diploid gynogenesis experiment. Phenotypically similar frogs occur among natural populations of the northern leopard frog. The animals have black eyes, dull gray dorsal skin between black spots, and a semi-transparent belly (Figure A-8) (Richards, Tartof, and Nace, 1969).

Where to Obtain R. pipiens

Although the leopard frog has the widest distribution of any anuran in North America with a range from Hudson's Bay in Canada to Mexico and Central America (Wright and Wright, 1949; Conant, 1975) (Figure A-5), it has been found that leopard frogs best suited for nuclear transplantation are obtained from the northern United States and certain areas of Canada. Little is known about the reproductive cycle of southern United States leopard frogs and, for this reason, it is not always possible to obtain females with fertilizable eggs.

Most dealers are located in Wisconsin or Vermont. Many biologists prefer to buy directly from frog dealers located near large natural populations of frogs rather than buy their animals from

Figure A-8. Melanoid variants of the leopard frog, *Rana pipiens*. The upper left frog is a melanoid Burnsi. The upper right is a spotted melanoid frog. Ordinary Burnsi and spotted frogs are in the lower part of the picture. Photograph courtesy of Dr. Christina Richards (from Richards, Tartof, and Nace, 1969.)

biological supply houses that in turn have had to obtain their animals from other collectors or dealers. Some large biological supply houses fill orders for frogs by sending their shipping label to a frog dealer who then takes frogs from his stock and sends them, with the supply house label, to the customer. Frog dealers are more likely to be able to provide frogs collected from a single locality (and thus eliminate the problem of possible nucleocytoplasmic incompatibilities) than are biological supply houses.

Seventeen commercial dealers of frogs are listed in DiBerardino (1967) and 59 sources that can supply not only frogs but a wide variety of living amphibians are listed by name and address in Appendix V (Nace and Rosen, 1978). The list in Appendix V is cross-referenced by species and location of supplier. *Animals for Research*, Ninth Edition, a 1975 publication of the Institute for Laboratory Animal Resources, National Academy of Sciences, lists suppliers of many common domestic laboratory animals and a diversity of invertebrates and vertebrates obtained from nature. Sources for many relatively exotic amphibian species are provided.

Storage of Frogs

Animals quarters are expensive to build and maintain. Fortunately, one may be parsimonious with research expenditures concerning the housing facilities for frogs. Ordinary household refrigerators costing about $250 are entirely adequate for holding northern *R. pipiens* until they are needed for experimentation. Obviously, walk-in cold rooms (4°C) are equally valuable if available.

The habitat of the leopard frog in the northern United States during winter is water under an ice sheet in a lake or stream with a temperature slightly above freezing (Emery, Berst, and Kodaira, 1972, McKinnell, 1973a). The cold-water frogs are torpid but are not considered to be in true hibernation. Natural conditions may be simulated by storing frogs in a refrigerator at 4° in glass or plastic containers partially filled with water. It is of convenience to the investigator that cold torpid frogs need not be fed. In my laboratory, I use 1- or 2-gallon, round glass aquaria (squat fish bowls) with galvanized wire cloth covers for storing frogs in the cold (Figure A-9). The wire cloth corners are bent in such a manner so that the covers stay on the bowls preventing the frogs from jumping out. Keeping frogs from jumping out is important because torpid laboratory frogs, just as frogs overwintering in an ice-covered lake, are capable of movement from time to time and the otherwise dormant animals sometimes climb out of their storage containers and arrive at their demise by desiccation. Five or six female frogs and six to eight male frogs (mature males are generally smaller than gravid females) can be placed in each glass container. Thus, a standard household refrigerator will hold up to 90 female and 120 male leopard frogs. The small glass aquaria have the virtue that they plus the water and frogs contained are of such little weight that they can be easily cared for. The aquaria water should be changed at least 3 times a week with pre-chilled water. Do not

Figure A-9. Northern leopard frogs to be used for experiments in reproduction must be kept refrigerated. Ninety to 120 frogs may be kept in artificial hibernation in a household refrigerator. Wire cloth tops prevent the frogs from climbing out of the glass containers. Frogs may be maintained as illustrated throughout the period of their natural hibernation.

subject frogs in simulated overwintering conditions to an abrupt temperature change.

What kind of water should be used for frog storage? The answer to that question is any kind of water which permits survival of frogs with eggs that will develop upon fertilization. Frog biologists have believed for years that salt added to storage water enhanced survival. Antibiotics seem to be needed because of the prevalence of "red leg." I currently store adult frogs in a 25% Ringer's

Solution (Appendix III) to which has been added 200 mg Neomycin sulfate (Biosol, Upjohn) per liter solution.

The northern leopard frog nuclear transplantation season extends from October through June. The season is limited by the time viable eggs and sperm can be obtained from adult frogs. Although ovulation can be induced in frogs obtained prior to the fall migration of frogs to the lakes and streams, I use only frogs caught at the time of entry into the lakes and frogs caught as they emerge from winter hibernation in the following April. Leopard frogs apparently do not feed during the time they are in the lakes. Thus, it is not necessary to provide food for frogs kept in simulated hibernation in the laboratory. Proper care for stored frogs thus becomes a matter of water temperature, frequent changes of non-toxic water, and keeping the storage containers clean.

A useful document published by the Institute of Laboratory Animal Resources (ILAR) of the National Academy of Sciences/ National Research Council provides much information on environmental control (water quality, lighting, and temperature), housing, and management of amphibians. The document, *Standards and Guidelines for the Breeding, Care, and Management of Laboratory Animals: Amphibians* (Nace et al., 1974), should be available in every laboratory attempting to do research with amphibians.

Water for Tadpole Rearing

The availability of non-toxic water for storing adult frogs and for rearing tadpoles is emerging as an increasingly important problem in many urban centers as ever greater quantities of herbicides and insecticides are found in natural bodies of water (Jacobson, 1967). It would seem that there is little reason to doubt the toxicity of modern pesticides to anurans. Adult frogs immersed for 30 days in very dilute preparations (less than 1 part per million) of halogenated hydrocarbon insecticides suffer from lethargy, convulsions, changes in skin color, reductions in red blood cell and total white blood cell counts, and decreased heart rate and respiration rate (Kaplan and Overpeck, 1964). Sixteen herbicides and insecticides were reported to be toxic to anuran tadpoles (Sanders, 1970).

Accordingly, considerable care should be exercised in selecting a source of natural water.

Great Bear Spring water is noted for its remarkable attributes as a culture medium for amphibian embryos by embryologists of the United States northeastern seaboard. Unhappily, this excellent water is unavailable in most of the United States. Whether or not to use natural water (spring, well, lake, or river water) or treated city tap water can only be ascertained empirically. It is suggested that a source of natural water be selected. The source should be in an area with minimal agricultural chemical runoff and optimally should be convenient to the laboratory. The capacity for the water to allow for normal development of frog eggs through metamorphosis should be tested. Mortality of normal-hatched embryos grown through to metamorphosis should not exceed 5%. Development of tadpoles should be normal. If lethality greater than 5% occurs or if abnormalities occur among the tadpoles to any substantial extent, it would be my judgment that treatment of tap water would be more economical and convenient than treatment of the natural water.

Testing the capacity of water to support normal development has become increasingly difficult during a time when many female frogs have eggs of poor quality. Several females should be ovulated and test fertilized. Only the egg batches with fertilization exceeding 99% should be retained for water testing.

There are several ways to treat tap water. Despite the fact that pesticides may be found in low concentration in drinking water, it has been my experience that there is nothing more complicated needed to treat some tap water than a few drops of a commercial dechlorinating agent (available at any pet shop) added to aquarium water that is aerated with a pump and a bubbling stone. Almost as simple to prepare as a substitute for natural water is charcoal-treated and filtered tap water (DiBerardino, 1967; Fankhauser, 1967). An inexpensive means of treating tap water when large volumes are not required is to condition the water for several days with bubbling air and many green water plants (Figure A-10). The conditioning procedure obviates the need for adding a dechlorinat-

Figure A-10. Glass aquaria set up for conditioning water for tadpole husbandry. The tanks contain green plants and air is bubbled through the water thus eliminating the need for dechlorinating chemicals.

ing agent to the water. Tadpoles have been reported to grow well in demineralized water that has been reconstituted by the addition of 0.45 g $CaSO_4$, 0.45 g $MgSO_4$, 0.72 g Na_2CO_3, and 0.3 g KCl per 15 liters water (Sanders, 1970). There is no substitute for an empirical test to determine which of the tap water treatment methods is best for tadpole rearing in a particular area.

I store frogs at $4°C$ from the time of receipt until they are used. Nace (1968) treats all newly received adult animals in water containing calcium hyopchlorite in a quantity sufficient to yield 6 parts per million chlorine for 10 minutes. Egg quality deteriorates if the females are warmed excessively, stored for more than 4 or 5 months, or if adequately pure water is not used.

Maladies of Frogs in the Laboratory

Amphibians in general and frogs in particular are subject to a wide spectrum of afflictions that render them unsuited for experimental purposes. It is not possible in this section to consider all of the ailments that bring about a moribund condition in anurans. More information may be obtained in the book by Reichenbach-Klinke and Elkan (1965).

Viruses. These agents will not be considered extensively because treatment for infected animals is unknown. However, several viruses have been studied and one or more may be of significance to the outcome of certain experiments. Two herpesviruses have been identified in the leopard frog. One herpesvirus is almost certainly the etiological agent of the Lucké renal adenocarcinoma (McKinnell, 1973a; Naegele, Granoff, and Darlington, 1974; and Chapter 4). The other herpesvirus was initially isolated from the urine of a tumor-bearing frog (Rafferty, 1965), is distinguishable from the Lucké tumor herpesvirus (Gravell, 1971), but has no known effect on the health of frogs. Frog virus 3 (FV3) (Lunger, 1966) is also known as the amphibian polyhedral cytoplasmic deoxyribovirus (PCDV). Infections of FV3 are lethal to embryos and larvae of leopard frogs (Tweedell and Granoff, 1968).

Bacterial Disease. "Red-leg" is a condition that results in some frogs having posterior legs and ventral surfaces red with inflammation and spotted with abscesses. However, the appropriateness of the name has been questioned because many frogs that die of "red-leg" never have reddened legs (Gibbs, 1973). Frogs with the disease have altered behavior. A sick frog manifests a "disinclination to move," lassitude, and a change in posture from an alert sitting-up appearance to the body and head assuming a horizontal position (Russell, 1898). The most obvious symptom became the name of the disease (Emerson and Norris, 1905), but it is not to be confused with a temporary redness that may result from drying and abrasions that occur when frogs are shipped. It occurs spontaneously among anurans in their natural habitat (Hunsaker and Potter, 1960) as well as in laboratory frogs. I have been told by frog dealers that, on occasion, they have had almost their entire stock of frogs obliterated by "red-leg."

The etiological agent of "red-leg" is thought to be *Aeromonas hydrophila.* The literature concerning the organism is complicated because of previously used synonyms that include *Bacillus ranicida*, *Bacillus hydrophilus fuscus*, *Bacterium hydrophilus fuscus*, *Bacil-*

lus hydrophilus, *Bacterium ranicida*, *Proteus hydrophilus* (Kulp and Borden, 1942).

The best way to control "red-leg" is to prevent its occurrence. Fresh water, cold temperature, and clean tanks in the storage facility are significant preventive measures. Immediate isolation of infected individuals is very important. It is well to clean thoroughly the storage tanks when a frog with "red-leg" is discovered. Keeping frogs in a dilute solution of salt has been recommended as a means of controlling "red-leg" epidemics (Reichenbach-Klinke and Elkan, 1965; Nace, 1968).

The variety of medications prescribed for "red-leg" attest to the difficulty of treatment of the disease. It has been reported that the lethal infection can be controlled by Neomycin sulfate (see section above on storage of frogs), copper sulfate treatment (Kaplan and Licht, 1955), chloromycetin (Smith, 1950), penicillin (Rafferty, 1962), tetracycline (Gibbs, 1963, 1973; Gibbs, Gibbs, and Van Dyck, 1966; Papermaster and Gralla, 1973), aureomycin (Reichenbach-Klinke and Elkan, 1965), and Pfizer's cosaterramycin, chloramphenicol, and sulfadiazine (Nace, 1968). It has been stated, and I enthusiastically concur, that bacterial infections can be avoided and even cured by maintaining animals in an optimal nutritional state with isolation and the use of appropriate antibiotics when necessary (Nace, 1968).

Certainly not nearly as important as "red-leg" but nevertheless of interest to anuran pathologists is the infrequent occurrence of tuberculosis in frogs (Elkan, 1960; Joiner and Abrams, 1967). Tuberculosis, "red-leg," and other diseases in an amphibian colony are discussed by Abrams (1969).

A Fungus Disease (Chromomycosis). A pigmented fungus pathogenic to *R. pipiens* was found in a group of 25 wild-caught frogs. The fungus is lethal to spontaneously infected and experimentally inoculated frogs and is a potentially serious problem in laboratory-housed frogs (Rush, Anver, and Beneke, 1974).

Protozoan Parasites. Several protozoan parasites have been isolated from blood and body tissues of leopard frogs. The parasites are of doubtful pathogenicity (Levine and Nye, 1976, 1977).

Induction of Ovulation

Ovulation in northern *R. pipiens* ordinarily occurs shortly after emergence of the frogs from the lakes in early spring. Ovulation may be induced earlier in the year from mid-October on through to the time of natural spring breeding by injection of freshly dissected pituitary glands following the procedure pioneered by Wolf (1929) and developed by Rugh (1934, 1962). There is a period (after spawning in the spring) during which no dosage of pituitary glands is adequate to induce ovulation simply because the frog lacks mature ova. Certain Mexican leopard frogs can be ovulated in July and August (Bagnara and Stackhouse, 1973).

Dissection of pituitary glands is a simple procedure. The head of a large female frog is removed. This is accomplished by a posterior cut made in the angle of the jaw followed by a transverse cut across the upper head posterior to the level of the tympanic membranes. The cut severs the spinal cord and after the opposite angle of the jaw has been cut, all of the head except for the lower jaw can be removed (Figure A-11).

The pituitary gland is removed from the upper part of the head by cuts made through the foramen magnum toward each eye. The flap of tissue and skull formed by the cuts in the roof of the mouth is reflected toward the anterior part of the head revealing the pituitary gland at the base of the brain. White endolymphatic tissue is found immediately lateral to the medially situated pituitary gland. The gland (Figure A-12) may be removed without damage by grasping the endolymphatic tissue with the points of jewelers' forceps and lifting. The endolymphatic tissue generally adheres to the pituitary gland and the gland is removed from the brain intact.

Pituitary glands obtained from large female frogs seem to have a superior potency (perhaps due to the large size of the gland) to those obtained from small female frogs or from male frogs. The dosage of pituitary glands sufficient to induce ovulation gradually decreases from about six pituitary glands in the fall to one gland during May or June. The dose of pituitary glands is reputed to be reduced slightly by disrupting the tissue in saline with a Ten-Broeck grinder prior to injection. The glands may be preserved for months

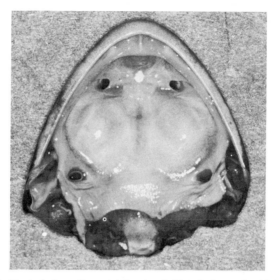

Figure A-11. Ventral view of severed frog head. The pituitary gland is located medially near the posterior margins of the eye bulges.

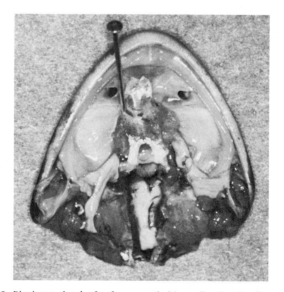

Figure A-12. Pituitary gland of a frog revealed by reflecting the floor of the skull anteriorly. White endolymphatic tissue borders the pituitary gland anteriorly and laterally. See text for description of ovulation induced by pituitary injection.

with little decrease in potency by freezing in a dilute saline solution (e.g., 10% Ringer's solution, Appendix III). The freezing compartment of a household refrigerator is adequate for this purpose.

Wild frogs, like other animals harvested from nature, are frequently infected with pathogens. Pituitary glands obtained from such animals may infect the female frog to be ovulated. Browder has described a method of extracting hormones from frog pituitary glands that eliminates the hazard of inadvertent transfer of pathogenic agents. He homogenizes glands from a number of female frogs in 70% ethyl alcohol and stores the mixture overnight in a refrigerator. The alcohol precipitates the ovulation-inducing hormones and presumably kills pathogens. The preparation is vortexed and then centrifuged. The supernatent is discarded and absolute ethyl alcohol is added to the precipitate. The preparation is centrifuged again and the supernatent is discarded. The extract is dried in vacuo and can be stored for up to a year. Ovulation is induced by injection of a water or saline suspension of the dried extract (Browder, 1975).

Leopard frogs should be considered to be a limited natural resource because of increased consumption in teaching and research and reduced natural supply (see discussion at beginning of this appendix). Hence, any method of frog conservation is of vital interest to the amphibian biologist. The progesterone procedure of Wright and Flathers (1961), if used with caution particularly near the time of natural ovulation (DiBerardino, 1967), affords a method of reducing frog consumption. One need inject only one pituitary gland in the early part of the season (in contrast to the usual six pituitary glands in the autumn) if that one pituitary gland is augmented with a dorsal lymph sac or intracoelomic injection of 0.5 to 5 mg of progesterone.

Pituitary- or progesterone-augmented pituitary-injected frogs are maintained at 18°C for 48 hours. Ova may be released prior to 48 hours post-injection but the early eggs generally do not fertilize or develop properly. Water should be changed at least twice daily after the frogs have been removed from the 4° simulated winter environment.

Fertilizable ova can be obtained from a pituitary-injected frog for several days to 1 week after ovulation. The frog is returned to

4°C after the 48 hour incubation period at 18°C following pituitary injection. The frog is allowed to come gradually to laboratory temperature (about 18°C) each time that eggs are to be obtained, and following the stripping of eggs, the frog is promptly returned to the refrigerator until needed again. Ovulated eggs in the reproductive tract of the female will not keep well if the female is allowed to remain at laboratory temperature for more than a few hours.

Fertilization

Sperm. A suspension of motile sperm may be prepared by dissection and teasing apart testes or by hormone-induced sperm release. Testes may be removed from a freshly pithed mature male frog. The gonads are rinsed with dilute saline solution (10% Ringer's solution, Appendix III) and then placed in a small glass dish (e.g., a 38 mm diameter Stender dish). The testes are then minced in about 10 ml of 10% Ringer's solution with jewelers' forceps. The mincing liberates mature sperm. The suspension of sperm is opalescent. A minimum of 15 minutes should be allowed for the sperm to become active.

A sperm suspension may be kept for up to 5 days if the container is covered and the preparation refrigerated between uses. To make use of stored sperm, first allow the liquid to equilibrate with room temperature, resuspend the sperm, which tend to settle to the bottom of the dish, by gently sucking in and out with a dropping pipette, and then employ the warmed and resuspended sperm as if it were a fresh preparation.

There may be a time when a particular male frog is needed as a sperm donor but the frog is rare or of exceptional value. Sperm may be obtained over a period of time by removal of but one testis through a small incision in an anesthetized frog. The incision may be clipped or sutured and the remaining testis saved for a future date. The method has the advantage of extending in time the life of a sperm source but the procedure suffers from the possibility that the frog may die due to the effects of the anesthesia or of infection and that there are only two testes. I believe that hormonal-induced sperm release is more advantageous than surgical

removal of testes because of the possibility of many sperm dona-
tions with minimal chance of death to the donor at each contribution.

Our method for sperm release mediated by hormone is as follows. A healthy mature male frog is selected as donor. It is allowed
to come to laboratory temperature for at least 1 hour. It is rinsed
with cool tap water, injected intraabdominally with 100 international units of commercial human chorionic gonadotropic hormone. The frog is then placed in a clean, transparent plastic box.
The frog's urine is collected after approximately 1 hour. The urine
is examined microscopically to confirm the presence of motile
sperm and is then used as a conventionally prepared sperm preparation (McKinnell, 1962b; McKinnell, Picciano, and Krieg, 1976).

Eggs. A gravid female frog (one that has been injected with pituitary glands or pituitary and progesterone at least 48 hours earlier and which has eggs located in the uteri) is held in one hand
and grasped in such a manner as to control both posterior legs
(Figure A-13). Gentle pressure sufficient to extrude eggs is maintained against the lower body while the frog is held in a vertical
position. The first 50 or 100 eggs extruded develop poorly (because of exposure to cloacal contents?) and are discarded. Ova are
then extruded onto a clean glass microscope slide (1 X 3 inches) in
a long row. If the frog has been stored in a refrigerator since it was
last used, it must of course be allowed to come to laboratory temperature before extrusion of eggs is attempted.

Fertilization and Early Stages. Sperm is pipetted onto the freshly extruded eggs. The sperm suspension is drawn back into the pipette and repipetted over the eggs every minute or so for a period
of 15 minutes. The excess sperm suspension is then discarded from
the slide and the freshly inseminated eggs are placed in a 100 mm
diameter Petri dish filled with dilute saline solution (10% Ringer's
solution, Appendix III).

The method described here results in nearly 100% fertilization
of eggs. Other methods such as placing large masses of eggs in volu-
minous dishes generally result in a lower percentage of fertilized
eggs. Even fertilization au naturel (by amplexus) in breeding ponds
seems to be less efficient than the method described here.

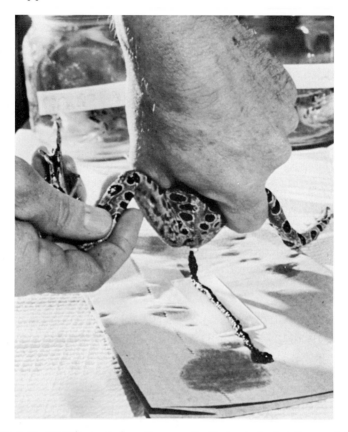

Figure A-13. Extruding eggs from the reproductive tract of a frog that was injected 48 hours previously with a pituitary gland-progesterone combination.

Fertilized eggs should be maintained between 15°C and 22°C
but they develop normally over an even wider temperature range.
Northern *R. pipiens* develop normally at 18°C and the time required to reach any particular embryonic stage can be predicted
with considerable precision by reference to Shumway's (1940)
stages of normal development (Appendix VI). Embryos are removed from the 10% Ringer's solution 1 day after fertilization and
placed in tadpole-rearing water or in a magnesium-fortified dilute
Holtfreter's solution (Brown, 1960, Appendix III).

Rearing Experimental Animals

Embryos (from fertilization to the onset of feeding) require little care. Animals should not be exposed to extremes in temperature. Crowding affects the growth of embryos and larvae (Richards, 1958; Gromko, Mason, and Smith-Gill, 1973; John and Fenster, 1975). Accordingly, each animal is reared in its individual dish (Figure A-14).

Sometimes it is desirable to rear the experimental animal to metamorphosis. This is a simple task if the tadpoles are properly cared for. Leopard frog larvae are vegetarians in nature but many kinds of laboratory food for tadpoles have been suggested. These include *Spirogyra*, liverwurst, Purina rabbit chow, oatmeal, and spinach (Merrell, 1963). Perhaps the most convenient diet is frozen spinach. It is readily obtainable from most grocers, is inexpensive, and requires boiling (for about 15 minutes) as its only preparation for serving. Boiled spinach can be kept for about 1 week under refrig-

Figure A-14. Feeding tadpoles in glass dishes. Tadpoles are reared one to a dish so that nuclear transfer data, chromosome constitution, and other information may be individually recorded and maintained.

eration. Tadpoles acquire kidney stones with a diet of spinach (Berns, 1965) but this in no way impedes their growth to metamorphosis. A more expensive diet that does not result in kidney stones is boiled romaine lettuce.

A compounded cuisine for larvae developed at Ann Arbor, Michigan, Amphibian Facility is probably far better nutritionally than a diet of lettuce or spinach. The Ann Arbor ration is composed of 250 g pulverized animal chow (Silver Cup trout chow, Purina No. 1 trout chow, or Purina rabbit chow), 20 g granular agar, 14 g unflavored gelatin, and 1,000 ml water cooked at 15 pounds pressure in the autoclave. The cooked mixture is poured into pans, frozen for storage, and thawed and sliced into ¼ inch thick slabs for feeding to larvae. The diet utilizing Purina No. 1 trout chow produced animals 3 times the weight of tadpoles fed lettuce at a substantial savings in money (Hirschfeld et al., 1970). In practice, larvae are started first on lettuce and then shifted to the new diet (Nace, personal communication).

One-quart, round aquaria (fish bowls) or small finger bowls are adequate for housing feeding tadpoles. Dishes should be washed regularly as a scum tends to form on the glass. Water should be changed often and only enough food added to last from day to day. Excessive food will hasten fouling of the water with bacterial growth.

Metamorphosis is a crisis in larval development. Care must be taken to ensure survival of the larvae through transformation. The tadpoles cease to feed at about the time one or both forelimbs protrude (Stage XX, Taylor and Kollros, 1946). Thus, food should no longer be presented to the larvae when the first forelimb becomes visible. During tail resorption, the bottom of the bowl should be lined with filter paper or some other suitable surface that allows for traction. The bowl should then be tipped so that part of the filter paper is under water and part is out of the water. The frog then has a choice of crawling over the filter paper to an aqueous or dry environment during transformation. The larval pigment pattern changes to that of an adult during the period of tail resorption. The time from fertilization to metamorphosis takes about 90 days. Transformation occurs faster with warm temperatures and slower at cold temperatures.

The investigator will be rewarded with large and healthy tadpoles that transform into vigorous froglets if the preceding procedure is followed with care. Despite these happy results, rearing tadpoles in individual aquaria provides an environment that is probably less than optimal. Two other procedures are known to provide slightly better results (as judged by the size of newly transformed froglets) but they require progressively greater expenditures of time, space, and effort.

Tadpoles will grow to relatively enormous size if they are reared in large (15 gallon) aquaria. The water should be aerated (air pumps and air-stones are available from pet shops). Abundant water plants should be provided which seem to be useful in providing cover for the animals. Larvae reared under these conditions result in young froglets substantially larger than those grown in small aquaria or finger bowls.

Perhaps the only thorough study of leopard frog husbandry methods ever made is that of the Amphibian Facility of the University of Michigan. High density culture is possible with superb results but the procedure requires substantial space, equipment, and a source of flowing treated water (Nace, Richards, and Sambuichi, 1966; Nace, 1968; see also Culley, Meyers, and Doucette, 1977).

Other helpful notes concerning the care of amphibian tadpoles may be found in Rugh (1962), Merrell (1963), DiBerardino (1967), and Nace et al. (1974).

The rearing of R. pipiens to sexual maturity in the laboratory has been reported only rarely (McKinnell, 1962b; Nace, Richards, and Sambuichi, 1966; Nace, 1968; Browder, 1968a). The rarity of the reports reflects the length of time and the care required in order to ensure survival of the experimental animals until they reach maturity. Husbandry of post-metamorphic frogs has been described recently by Culley (1976). A sexually mature nuclear transplant frog with two of its progeny is illustrated in Figure 2-3. The frog was reared in a large aqua-terrarium that had a number of lush green plants and both dry and wet areas for its occupants. It was fed a variety of creatures that included *Drosophila melanogaster*, *Tenebrio* larvae, and earthworms. It was also forced fed for a time with liver dipped in a mixture of bone meal and cod liver oil. The frog's progeny vouched for its sexual ripening (McKinnell, 1962b).

Possible Health Hazards Associated
with Frog Husbandry

Zoonoses are animal diseases transmissible to man. Leopard frogs and other anurans have ailments that also afflict man. The peril to humans by contact with frogs is considered so minimal that experiments with frogs constitute part of the biology curriculum of almost all colleges and universities. However, individuals with long-term research commitments contemplating prolonged exposure to large numbers of animals should be informed of biological hazards so that preventive measures may be taken to minimize risk. The significance to man of the following pathogenic agents is unknown but information concerning them is included below to make biological workers aware of possible hazards. Most of the following pathogens have been described in species other than *R. pipiens*. They are listed because several types of frogs may be similarly infected and because experimenters frequently use species other than the leopard frog.

Viruses. Western equine encephalomyelitis virus causes an acute inflammatory disease in the central nervous system of horses and man. It has been found in leopard frogs in Saskatchewan (Spalatin et al., 1964).

An oncogenic virus that is the etiological agent of a prevalent renal tumor of leopard frogs is described in Chapter 4 of this book. There is emerging considerable concern for the safety of individuals who work with oncogenic viruses (Wade, 1973; Hellman, Oxman, and Pollack, 1973). Laboratory personnel who work with oncogenic viruses must be trained in microbiology and biohazard control and containment. Children, individuals on immunosuppressive drugs, and pregnant staff members (Chesterman, 1967) are excluded from the laboratory. Eating, drinking, and smoking is prohibited in the laboratory where oncogenic viruses are studied. There is no way in this brief space to deal adequately with methodology for biohazard management. The prohibitions cited above are listed to suggest that unusual precautions are mandatory. The new investigator must seek appropriate guidance from the current literature of viral oncology.

Bacteria. Leptospirosis icterohaemorrhagica (Weil's disease) is caused by a spirochete that may be found in laboratory animals. The disease is characterized by an early high fever followed by nausea, headache, and muscular pains. Jaundice is sometimes observed. The illness may be fatal. Leptospires have been isolated from natural bodies of water in areas where frogs are collected for commercial purposes (Diesch and McCulloch, 1966) and from the kidneys of frogs (Ellinghausen, 1968). Infection of man by leptospires is believed to occur through breaks in the skin, through sodden skin, and through oral, nasal, and conjunctival mucous membranes (McCulloch, Braun, and Robinson, 1962). Because Weil's disease may be fatal and because some frogs are known to harbor a leptospire (which, fortunately, may be limited in its epidemiological importance, Babudieri, 1972), it seems reasonable and prudent to instruct laboratory personnel to wear rubber gloves if they have cuts or cracks on their hands and to wash hands carefully prior to touching mucous membranes after working with frogs.

Thirteen serotypes of *Salmonella* were isolated from 31 of 164 frogs imported into France as food (Pantaleon and Rosset, 1964). Symptomless anurans have been shown to carry salmonella serotypes implicated in human salmonellosis (Kourany, Myers, and Schneider, 1970). Tuberculosis infections of amphibia include one species of *Mycobacterium* that infects both amphibians and man, although probably an opportunist in both. *M. xenopei* has been found in *X. laevis* and is not uncommon in human beings, mostly in the London area of England (Marks and Schwabacher, 1965). *Aeromonas hydrophila*, the etiological agent for red leg in frogs, may be pathogenic to man as well (Graevenitz and Mensch, 1968).

Frogs and toads may be infected with the anthrax bacterium, which although primarily an agent of animal diseases, can cause malignant pustule in humans. People generally aquire the disease from animals through a lesion or scratch in the skin (Hull, 1955). *Brucella*, the agent that causes undulent fever, has been identified in *R. esculenta* and *B. vulgaris* (Galouzo and Rementsova, 1956).

R. catesbeiana is a natural host for *Pasteurella tularensis* and *R. esculenta* is reported to be susceptible to the agent of tularemia (Van der Hoeden, 1964; Pavlovsky, 1966).

Fungi. No fungus disease of man is known to be aquired from amphibians but a fungal infection of man is caused by *Basidiobolus* in tropical Africa. The fungus also lives as a saprophyte in the gut of frogs and frog droppings could serve as a source of infection (Burkitt, Wilson, and Jelliffe, 1964; see also Rush, Anver, and Beneke, 1974).

Helminths. A worm in the retina of the human eye was tentatively identified as *Alaria*. It was believed to have been inadvertently placed in the eye with contaminated hands after skinning freshly killed frogs (Shea et al., 1973).

Orientals sometimes treat infections by application of poultices of fresh frog flesh (e.g., uncooked muscle from *R. nigromaculata, R. limnocharis,* and *Hyla coeruela*). Larval parasites migrate from the frog tissue to the warm human lesion, enter, and cause sparganosis (Van der Hoeden, 1964; Manson-Bahr, 1966).

It would seem that the diversity of pathogenic agents, ranging from viruses to worms, found in several species of frogs would be such as to discourage the keeping of anurans as pets and to encourage laboratory precautions to protect the health of workers.

Cloning Procedures

A. The Fashioning of Microimplements

Microneedles and micropipettes suitable for nuclear transplantation are not available commercially and the skill necessary to fabricate these implements must be learned by the person who desires to engage in the craft of nuclear transplantation.

Microneedles

Very fine glass needles are required for activation (pricking) and enucleation (removal of the second meiotic division spindle and chromosomes) of the recipient egg. Although enucleation of the host egg can be accomplished by laser irradiation (Appendix II, Part b), a glass needle is nevertheless required to activate the egg artificially prior to enucleation. Thus, there is a continuing need for suitable microneedles regardless of the procedure used for enucleation. Glass microneedles are useful also for dissection of embryos to be used as donors of nuclei and for other procedures in experimental embryology.

While it is true that an ordinary glass rod can be pulled to a relatively small diameter and that glass needles can be drawn from this rod with the aid of a small gas burner, the taper of the resulting microneedles generally varies. The lack of uniform microneedles is frustrating to the nuclear transplanter. Attempted enucleation with too thin a microneedle results in a microneedle tip bent to resemble a light-weight fly rod with a 10-pound largemouth bass on the line. It is virtually impossible to control the tip of a severely bent microneedle. A stubby needle is of little more value. Because of these reasons, a mechanical needle puller of some variety is desirable to assure uniformity and proper taper.

Several kinds of needle pullers are available commercially. A needle-drawing apparatus for micrurgical work is available from Carl Zeiss Jena as is a vertical pipette puller (model 700c) from David Kopf Instruments. The E. Leitz, Brinkman, and Rachele companies also make devices for needle and pipette pulling and are illustrated in El-Badry (1963).

I use the micropipette puller designed by Dr. Luzern G. Livingston[1] of Swarthmore College with entirely satisfactory results (Figure A-15; see also Chambers and Chambers, 1961, figure A.4; King, 1967, figures 2 and 3). The Livingston micropipette puller is similar in operation to the Du Bois (E. Leitz) puller. The micropipette holders move upward and apart in an arc as the glass tubing or rod softens. The instrument is designed so that glass is under relatively constant tension as it is pulled. The heating element is made of #27 platinum wire mounted in mica plates, which are in turn placed in a bakelite frame with 1 mm slots at both ends. The slots are important in alignment of the heating element with the pipette holders. Proper alignment is crucial to the well being of the heating element because if melted glass touches the platinum heating wire, the glass will stick to the wire and pull the wire out of its proper position.

Micropipettes and microneedles of different shape and flexibility can be fashioned by varying the following: (1) intensity of heat, which is controlled by the spacing of the platinum wires and by the setting of the variable voltage transformer; (2) duration of heat, affected by both the ratio of horizontal to vertical movement (the arc of pull can be controlled by where the gear wheels are set when the glass is clamped into position) and by the depth of the glass in the heating element (the heating element can be lowered or raised on its column), and (3) melting point of the glass stock used. Adequate instructions for the fashioning of microneedles and micro-

Figure A-15. Livingston micropipette puller and transformer. The intensity of heat, extent of time the glass is exposed to heat, and tension exerted on the glass can be varied to produce a variety of kinds of microneedles and micropipettes. The glass rod or tubing is fastened into position by clamps (A) that position the glass stock within the heating element (B). Intensity of heat is varied by the transformer at right. The tension or pull is varied by the extent to which the spring adjustment (C) is extended to the left and also by how deep into the heating element the glass stock is placed. The height of the heating element can be adjusted and this varies the taper of the microneedle and micropipette.

pipettes are included with the Livingston instrument when it is purchased. It has been my experience that considerable experimentation is necessary in order to obtain microneedles of proper shape. Therefore, I record for future use the settings that yield useful microneedles and micropipettes. It should be noted that the empirically derived settings may become useless if the machine becomes dirty or if it is not oiled regularly. It is suggested that the micropipette puller be kept free of dust with a plastic cover and that its use be restricted to only a few well-qualified operators.

What follows is an example of how a particular set of conditions, which allowed for the manufacture of microneedles for activation and enucleation, were ascertained on the Livingston instrument. Glass rod 1 mm diameter (± .025 mm)[2] was used throughout. The wheel setting was constant at 1½ (the wheel setting is read when the glass rod is clamped securely in place; the markings are etched on the lower part of the left wheel). The heating element wires were aligned with the die to set the wire gap maximally. The height of the heater element was set at 5 (as read when the glass stock has been clamped into position). The heater is raised or lowered by twisting the knurled nut on the heating element pillar. A suitable temperature was selected by pulling needles with different settings on the variable voltage transformer. With the wheel at 1½, the heater column at 6, and the transformer at 60, rigid, short needles were formed. Longer needles were obtained with the transformer at 70 and still longer needles with the transformer set at 80. It was judged that the most useful temperature for microneedle formation was with the transformer at 80. The final shape of the microneedle was then determined by varying the height of the heater element. Long, hairlike microneedles were formed with the heater element at 5; stiffer, somewhat shorter needles were formed at 4; and needles too short to use were forged at heights of 3 or less. The combination of wheel set at 1½, transformer set at 80, and heater element set at 4 was judged to result in suitable microneedles; accordingly, a number of essentially identical needles were fabricated.

A microneedle useful for both activation and enucleation should be sharp and thin enough to penetrate the jelly coat of an egg and prick the egg cortex with little disruption to the latter but it should be sufficiently rigid so that it can be inserted under the maternal chromosomes and spindle and can lift them away to form an exudate from the ovum. A microneedle that is too thin is useless because it bends so much that enucleation becomes impossible. The tip of a microneedle suitable for both activation and enucleation is illustrated (Figure A-16).

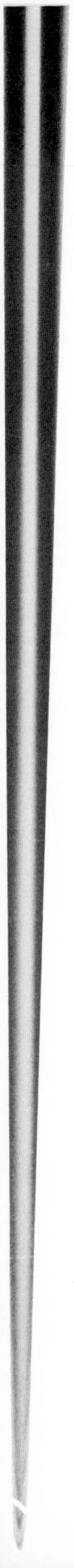

Figure A-16. Photomicrograph of a microneedle tip suitable for activation and enucleation of frog eggs. The taper of the microneedle is such that it will be rigid enough to form an exudate from the egg containing the meiotic spindle and chromosomes but is slender and sharp enough not to seriously damage the egg cortex with activation.

Micropipettes for Transplantation

A well-constructed micropipette is absolutely essential for nuclear transplantation.[3] Perhaps nothing is more disrupting to the happiness of a nuclear transplanter than a micropipette that is too dull and that stretches, deforms, and indents the cortex without penetrating the surface of the recipient egg. Criteria for selecting a good micropipette include: (a) proper diameter of orifice, (b) nearly uniform diameter near the orifice, and (c) a beveled tip that terminates in a microscopically sharp point.

Precision drawn 1 mm (± .01) Pyrex capillary tubing for the forging of glass micropipettes may be obtained from the same supplier that provides glass rod used in the manufacture of microneedles.

The first step in the manufacture of a micropipette is the drawing out of a piece of glass tubing in the Livingston micropipette puller (Figures A-15 and A-17). The micropipette just posterior to the pointed tip should be of a diameter needed for the cell type under consideration. The micropipette diameter should be somewhat smaller than the diameter of the cell that is to serve as a nuclear donor. A pipette diameter of 37 μ has been found to be suitable for animal hemisphere late blastula cells as well as for endoderm cells from late gastrulae.

Micropipettes of sizes other than 37 μ seem to be less suitable for transplantation of late blastula nuclei. Pipettes smaller than 37 μ require the selection of correspondingly smaller donor cells that have less protective cytoplasm. Nuclear distortion probably occurs within the smaller micropipette. Yolk loss from the recipient egg is frequently a problem with micropipettes larger than 37 μ. The implanted nucleus may be lost from the recipient egg as yolk gushes forth from the incision made by a large pipette. Because of these reasons, a micropipette of 37 μ diameter seems to be of optimal size.

The micropipette should be of approximately uniform diameter for several mm near its tip. The transplantation fluid meniscus and

Figure A-17. Livingston micropipette puller with glass tubing (*A*) positioned within the heating element (*B*) in preparation for the forging of a micropipette.

the cell in the tip of the micropipette tend to creep up a micropipette of uneven taper resulting in poor control during transplantation. The adjustments in heat intensity and duration of heat of the micropipette puller for suitable micropipettes, like microneedle settings, are ascertained empirically.

After the micropipette has been pulled on the Livingston instrument, the next step in the manufacture of the micropipette is the formation of a beveled and sharpened point. I use the Sensaur de Fonbrune microforge (Model MF-67) for this procedure[4] (Figures A-18 and A-19). The microforge is an instrument designed by Dr. Pierre de Fonbrune at the Pasteur Institue (de Fonbrune, 1932, 1949) and has the capacity to aid in the fabrication of a wide variety of microtools used in micromanipulative and microsurgical procedures. The apparatus allows for separate manipulation of the platinum-iridium heating filament and an instrument holder viewed with an inclined stereoscopic microscope. The microforge has a

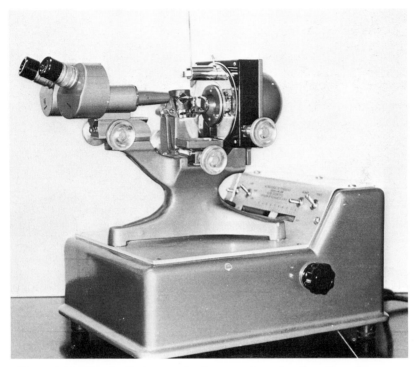

Figure A-18. The de Fonbrune microforge consists of a dissecting microscope mounted with its optical axis in the horizontal plane, a heating element, cooling jets of air, and devices for the micromanipulation of glass rods and tubes. See text for description of its operation.

blower that directs jets of air near the working area of the heating filament such that high temperature is confined to one small area of the glass instrument being manufactured. With this arrangement of cooling air near the hot filament, a micropipette is not fused (i.e., the lumen remains patent) by the temperature required to fashion a sharp tip.

An unfinished micropipette as it appears through the microscope of the de Fonbrune microforge is illustrated in Figure A-20. Note that the diameter of the tubing is almost uniform in the area near the platinum-iridium filament.

The micropipette after it and its microtool holder have been placed in the microforge is then moved against the cold platinum-

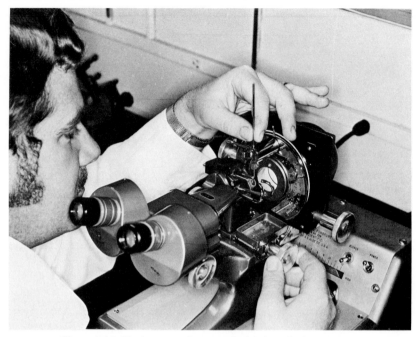

Figure A-19. Worker removing a newly fabricated micropipette
from the de Fonbrune microforge.

iridium filament in such a manner that the tip bends (Figure A-21)
and then breaks with a smooth bevel (Figure A-22). Finesse is required for judging where to bend the micropipette. A micropipette
with an aperture too small is formed by brushing the glass near its
pointed tip against the filament. In contrast, if the micropipette is
bent too distal from its end, its opening will be excessively large.

An alternative method for breaking the tip of a micropipette is
illustrated in King (1966). King's method involves fusing the micropipette to the heated filament of the microforge. The filament contracts and breaks the pipette at the point of fusion with the filament when the electricity is turned off. The method described by
King may be modified so that precision-made micropipettes can be
produced rapidly and with considerable ease. The modification involves bending and breaking the unfinished micropipette with a
hot filament. Glass tubing previously pulled in the Livingston mi-

cropipette puller is placed in the instrument holder of the de Fonbrune microforge. The micropipette is placed a short distance to the right of the cold filament as viewed through the dissecting microscope. The blower is turned on and the heating element is also activated such that the filament has a barely discernable red glow. The filament is moved while hot so that it barely touches the micropipette at a point where the inside diameter of the glass is approximately 37 μ. The distal free end of the micropipette will then soften at the point of contact with the filament and bend toward the heating element. When the bent distal portion of the micropipette becomes about 30° away from the axis of the upper portion of the micropipette, the current to the heating filament is cut off. The heating element abruptly contracts and when it does so, it breaks the micropipette (the distal end of which adheres to the heating filament) with a perfect bevel. All that remains to be done to the beveled micropipette is to put a fine point on its tip as described below.

With some experience, the microforge operator can learn to obtain micropipettes with tips of a smooth bevel and orifices of appropriate diameter with either procedure. All that remains to be accomplished in the forging of the micropipette is the sharpening of the drawn and beveled tip. The micropipette is sharpened by racking it down slowly on the heated filament of the microforge with the blowers thrusting cool air near the heated filament. The tip of the glass is touched to the hot wire and then drawn away. This results in a sharpened or spear tip of the micropipette. The ideal tip is sharp enough to allow for easy penetration of the recipient egg but not so long and slender as to bend or break (Figures A-23 and A-24).

The diameter of micropipettes is accurately measured by use of an ocular micrometer in a compound microscope. I use a magnification of 125 X to examine and measure newly fashioned micropipettes. A glass microscope slide with a slab of cork, 25 mm in diameter and 4 or 5 mm thick, with a groove on its surface parallel to the long dimension of the slide may be used for holding the micropipette while it is being viewed with a compound microscope. The micropipette is placed on the cork with 10 to 15 mm of its tip extending beyond the cork. With this procedure, the tip may

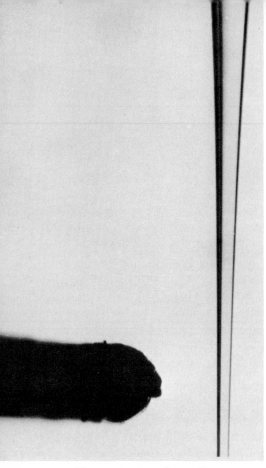 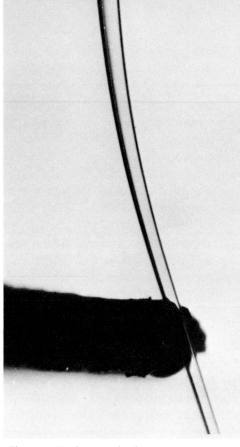

Figure A-20. Part of an unfinished micropipette as viewed through the microscope of a de Fonbrune microforge. The dark structure in the lower left is the platinum-iridium heating filament of the microforge. Note that the micropipette tapers little in the area opposite the filament.

Figure A-21. Photograph of an unfinished micropipette as it is bent against the cold heating filament of a de Fonbrune microforge. Note that the glass micropipette can withstand substantial bending without breaking. Additional bending will result in a break above the area of contact with the filament.

be examined with the microscope without touching the glass slide. A finished micropipette is so delicate that even the slightest touch of its tip to a solid surface will damage it. The slide and cork slab allow for the micropipette to be positioned under the microscope objective with a conventional mechanical stage. The micropipette tip stays in focus and does not vibrate when placed on the cork and slide (Figure A-25).

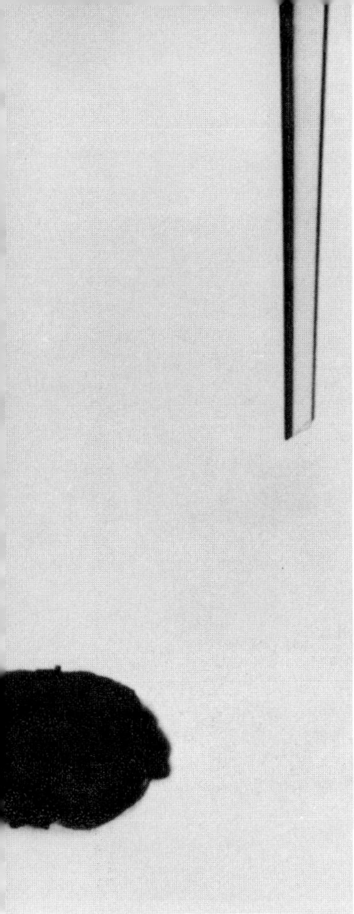

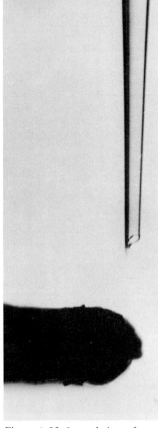

Figure A-22. Lateral view of an unfinished micropipette that has been broken by pressure against the cold filament of a de Fonbrune microforge. Note that the break has resulted in an aperture that has a bevel. The diameter of the tubing near the aperture is 37 μ. The point of the micropipette is not yet sharpened.

Figure A-23. Lateral view of a micropipette with the following desirable characteristics: the aperture is 37 μ, which is suitable for animal hemisphere blastula cells; the bevel of the aperture is smooth; the micropipette has been sharpened to a fine point by touching its tip and slowing drawing it away from the heated platinum-iridium filament; and the lumen of the micropipette is of nearly uniform diameter near its tip.

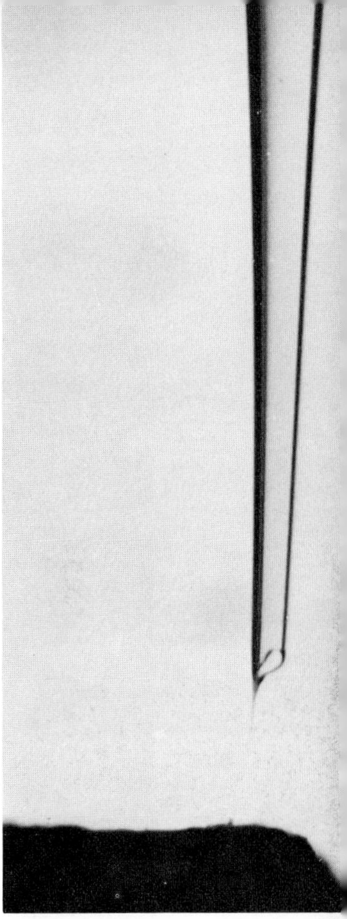

Figure A-24. Lateral view of another finished micropipette with 37 μ diameter aperture. It has many of the desirable features of the micropipette shown in Figure A-23. Note, however, that there is an irregularity to the bevel which may or may not interfere with insertion into the egg cortex and the sharpened tip is excessively long. The very fine glass at the point will probably break during the manipulation of the micropipette attendant to the picking up of the donor cell. If it does not break at that time, it most likely will break upon insertion into the egg cortex. Frequently, the usefulness of the micropipette is unaffected when the extreme tip is broken.

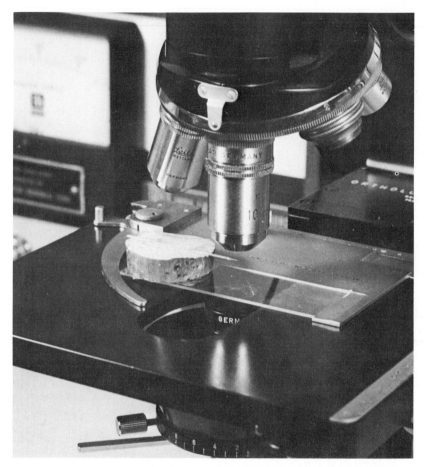

Figure A-25. Simple glass microscope slide with section of grooved cork glued to it for the examination of micropipettes. Tiny irregularities in the glass tool may adversely affect the nuclear transfer operation. The irregularities can be detected without damage to the micropipette by this method.

Care of Microimplements

Needles and pipettes may be stored conveniently in wooden racks covered by inverted beakers. Storage in this manner protects against breakage and minimizes dust accumulation. The wooden racks that I use in my laboratory are turned on a lathe with an inner raised portion about $2\frac{7}{8}''$ diameter that will just accommodate a 400 ml beaker. Holes $1/16''$ diameter and approximately $3/8''$ deep are drilled in the wood to accommodate 36 microneedles or micropipettes (Figure A-26). A high pressure hose is used to blow away small wood chips that remain after drilling so that they do not become lodged in the micropipettes.

Clean micropipettes are absolutely essential in the nuclear transplantation procedure. Cleaning may be accomplished by drawing a hot mixture of sulfuric acid and nitric acid in and out of the pipette several times followed by several distilled-water rinses. The hot acid cleaning is performed in a chemical hood. Microneedles used for activation and enucleation can be cleaned in the same hot acid mixture. Following acid cleaning the pH of the pipette should be checked by drawing into the pipette brom-thymol blue (pH 7), then rinsing with distilled water again.

Figure A-26. Micropipette and microneedle storage racks. The inverted 400 ml beakers protect the glass tools from dust and from breakage and thus safeguard the unwary laboratory worker from being injured by many microscopically fine glass fragments.

A good micropipette is a tool to be cherished. It may be used repeatedly provided it is cleaned properly between uses. Jelly and a film of yolky cytoplasm tend to accumulate on a pipette during an experiment. The film is easily removed immediately following use of the pipette with a 10% solution of commercial bleach, such as Chlorox (= 0.5% sodium hypochlorite). Dried jelly and cytoplasm are substantially more difficult to remove than the moist fresh material. The hot acid cleaning follows the bleach.

Some journeymen of the nuclear transfer craft find that constructing micropipettes is simple and cleaning dirty micropipettes is time consuming and intellectually unrewarding. These individuals construct, use, and discard micropipettes and eschew the laundering process.

B. Nuclear Transplantation

The term "nuclear transplantation" is conceptually correct but it is misleading from a procedural point of view. What is accomplished is the placing of a broken cell (i.e., a nucleus with practically all of its cytoplasm intact but with a ruptured plasma membrane) into a previously activated and enucleated virgin egg. The steps necessary to accomplish nuclear transplantation in the northern leopard frog, *Rana pipiens*, will be described in this section. Additional notes will be provided toward the end of Part B for the procedure as it applies to the African clawed frog, *Xenopus laevis*, and to the Mexican axolotl, *Ambystoma mexicanum*.

Preparation of Recipient Egg Cytoplasm

Conventional Manual Enucleation. Eggs of *R. pipiens* can be activated and enucleated with a hand-held needle holder with glass needle. Freshly extruded ova are activated by pricking with a clean glass microneedle and enucleation is accomplished by a modification of a method originally devised as a means of producing androgenetic haploid embryos (Porter, 1939). Activation and enucleation in *R. pipiens* is best accomplished with eggs obtained from recently ovulated females. I suggest the use of fresh eggs in spite of the fact that I am aware of a report that suggests over-ripe eggs (i.e., cloacal eggs obtained from females kept 4 to 5 days at 18° to 22°C) do not differ in their capacity for cleavage from normal fresh eggs similarly transplanted (Sambuichi, 1957b). It should be noted that activated *unfertilized* eggs are enucleated in this procedure—a zygote nucleus is not removed, as is sometimes incorrectly described, for these operations. There are two perfectly respectable reasons why an embryologist working with leopard frogs would choose *not* to enucleate a fertilized egg. First, the egg that is destined to serve as a recipient of a transplanted nucleus would be contaminated with thousands of live sperm are utterly competent to initiate development in the absence of another nucleus. Additionally, sperm nuclei are competent to fuse with a transplanted somatic nucleus. Consequently, sperm sterility is absolutely necessary in the nuclear transfer experimental procedure. The other reason that a zygote nucleus is not removed is that the zygote nucleus is situated deep in the interior of the fertilized egg in decided contrast to the superficial location of the meiotic chromosomes and spindle. The meiotic genetic material is accessible to treatment by irradiation or surgical removal, whereas the zygote nucleus is denied to the conventional manipulations of the microsurgeon.

Eggs are obtained from a frog previously injected with pituitary glands alone or pituitary glands with progesterone (Appendix I). The frog is allowed to come to room temperature if it has been refrigerated. Eggs are expressed from the gravid female onto a 3 × 1 inch glass microscope slide as described for fertilization. The slide containing the eggs is then placed in a 100 mm diameter Petri dish containing 10% Ringer's solution or 10% Magnesium-fortified Holtfreter's solution (Appendix III). The Petri dish with its eggs and solution are placed under low-power magnification of a stereoscopic microscope with two good microscope lamps (American Optical Universal microscope lamps or equivalent lamps).

A previously prepared microneedle is inserted into a needle holder. Each egg is carefully pricked with the hand-held glass microneedle in the pigmented animal hemisphere near the equator with the tip of the microneedle penetrating the jelly and cortical cytoplasm. Pricking results in parthenogenetic activation of the egg.

The egg will rotate within its vitelline and jelly membranes in 5 to 15 minutes at 18°C and the second polar body will start to form. The microneedle tip is directed to the equatorial area of the egg during activation in order not to disturb the preliminary events that lead to polar body formation. The polar location of the meiotic spindle is revealed by an accumulation of pigment granules that cluster about the spindle and which appear externally as a black dot (Porter, 1939; Subtelny and Bradt, 1961; Briggs and King, 1953; King, 1967).

The activated egg may be enucleated by placing the tip of a glass microneedle on the proximal side of the black dot and inserting the microneedle into the cytoplasmic cortex under the dot and through the cortex to the exterior again slightly distal to the black dot. The microneedle is then lifted away from the egg slowly and held in position until a cytoplasmic exudate is formed (Figure A-27).

Sometimes many of the ova are upside down or found on the slide in a position such that the animal hemisphere is obscured. Enucleation is thus difficult. A means of obtaining a greater proportion of ova with animal hemisphere up follows. A small finger bowl containing a 3″ × 1″ slide is filled with 10% Ringer's solution (Appendix III) or distilled water. Eggs are extruded at the water surface while the gravid female is briskly moved about. When properly done, individual ova (i.e., ova not in a "string") slowly descend through the liquid and generally reach the glass slide in an upright position. The slide and its eggs are then placed in a Petri dish for activation and enucleation.

It is extremely frustrating to attempt manual enucleation if the eggs do not adhere to the glass slide when it is flooded with dilute saline. Although the jelly coat of eggs will sometimes cling to a clean glass microscope slide, they do not stick frequently enough for ordinary microscope slides to serve satisfactorily for this purpose. I use a Pettus slide (devised by Dr. David Pettus of Colorado State University, Fort Collins, Colorado) in place of a conventional glass slide. A Pettus slide is a 3 × 1 inch glass microscope slide that has been rubbed with a Carborundum[R] abrasive stone until it has a frosted appearance.[5] Eggs extruded on such a slide and flooded with a dilute saline solution do not pull loose during the

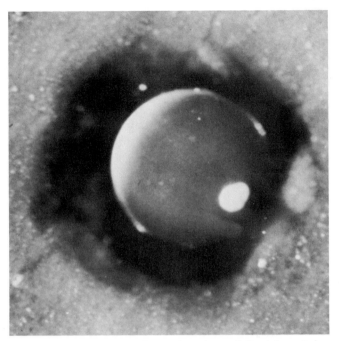

Figure A-27. Activated and manually enucleated egg of *R. pipiens*. A depression in the agar substrate holds the dejellied ovum, which is immersed with transplantation medium. The maternal chromosomes are contained in a cytoplasmic exudate that appears as a white sphere on the animal hemisphere.

manipulations that move the eggs about with activation and enucleation.

A means of assessing the effectiveness of the enucleation process is described in Briggs and King (1952). The black dot associated with second polar body formation is removed following conventional insemination with a fresh sperm suspension. Androgenetic haploid embryos develop if the maternal chromosomes are successfully removed; diploid embryos are a consequence of faulty enucleation procedure. Briggs and King reported that of 358 enucleation operations of inseminated eggs, 337 developed to embryonic stages. All of the 337 were haploid.

A hand with a tightly grasped needle holder containing a microneedle tends to shake. Apparent tremor may be considerable when viewed under the dissecting microscope. It has been my experience that normal tremor is controlled to tolerable limits by resting both forearms on the work bench with the microscope and chair heights adjusted for maximal comfort. A brief period of caffeine abstention may be beneficial in minimizing excessive shaking.

Enucleation by Means of Laser Irradiation. An alternative method of enucleation for eggs of *R. pipiens* has been developed utilizing laser radiation. Eggs to be laser enucleated are activated parthenogenetically with a clean glass microneedle in the conventional way. The maternal meiotic apparatus is then inactivated by directing a microbeam ruby laser at the spindle and chromosomes. The pigment granules that are clustered around the second meiotic spindle apparently absorb the laser radiation and the radiant energy is converted to heat. A small coagulum on the surface of the egg (the size of which is a function of the energy input to the laser) is the visible manifestation of laser irradiation.

The source of laser radiation that I use in my laboratory is the Hadron/TRG Model 513 Biolaser[6] system installed on a Leitz Ortholux microscope (Figure A-28). The apparatus emits pulses of high-intensity monochromatic laser radiation in the form of an ultra-narrow coherent light beam. The instrument has a pulse length of 150 μ seconds with a wave length of 6,943 Å.

The laser beam is aligned with the microscope so that there is coincidence of the cross hairs in an ocular reticle with the focused beam at the stage of the microscope. The target area (in the present case the meiotic black dot of an activated egg) is then positioned with respect to the cross hairs of the ocular reticle. The laser is triggered when the black dot and the cross hairs are coincident. Laser irradiation is exceedingly hazardous to the human eye, and the instrument is installed with a safety device that prevents triggering the laser while viewing the object to be irradiated. Simultaneous viewing and triggering would direct the laser beam into the observer's eyes but this is not possible with the safety interlock. The operator should be advised to turn his back to the instrument when it is triggered because hazardous specular reflection is possible from the slide and the material being irradiated.

The Hadron/TRG Model 513 Biolaser system may be fired at the rate of one pulse per minute. At this rate, 10 to 20 eggs can be

Figure A-28. Leitz Ortholux microscope fitted with a Hadron microbeam ruby laser for enucleation of activated eggs of *Rana pipiens*. The maternal meiotic apparatus of unfertilized eggs is aligned with cross hairs in the optical system of the microscope and irradiated with a laser pulse. (From McKinnell, 1973b.)

enucleated from a single batch of activated eggs. This number is generally adequate for a nuclear transplantation experiment. The rate of firing can be doubled if desired by installation of a cooling hose to the laser head. Pressurized nitrogen is recommended for laser cooling.

The diameter of effective irradiation is 120 μm utilizing 120 joules energy input and the 3.5 $\times$ scanning lens of the Leitz Ortholux microscope. Jelly and lightly pigmented cytoplasm seem to be not as sensitive to ruby laser irradiation as is the black dot. We have found that laser irradiation with the system described here is effective in inactivating the maternal meiotic apparatus without causing developmentally significant cytoplasmic damage (McKinnell, Mims, and Reed, 1969; Ellinger, King, and McKinnell, 1975).

The principal benefit of laser inactivation of the maternal meiotic chromosomes versus manual enucleation is the lack of cytoplasmic loss by the former method.

Concerning the Removal of Jelly from Eggs. Removal of outer jelly envelope follows the enucleation procedure. Stainless steel jewelers' forceps are suitable for this purpose. The jelly is gently torn away from the enucleated egg in such a manner that the cytoplasm is minimally distorted in the process. One operator can activate, enucleate, and remove the jelly from 20 eggs in about 45 minutes or so. Preparation of more than 20 eggs does not seem to be profitable because of the lapse of time from activation to nuclear transfer. The cleavage of the recipient egg is associated with the time of activation and not the time of nuclear transplantation. Accordingly, nuclear implantation must take place during the relatively brief interval while the recipient egg cytoplasm is competent to interact with the inserted nucleus.

Eggs that have been "dejellied" as described, in fact, still have a thin inner jelly layer as well as the vitelline membrane. The thin jelly layer tends to enlarge as the eggs sit in dilute saline. The inner jelly coat essentially can be eliminated by a 20-second treatment with a mixture of 2% cysteine-HCl, 0.2% papain, and 0.2% alpha-chymotrypsin (Hennen, 1973). The treatment also softens the vitelline membrane, which enhances ease of micropipette entry into the recipient egg.

The enucleated and jellyless eggs should be placed in full-strength operating solution prior to nuclear insertion. They are kept in full-strength solution so that the transplantation medium is not diluted when the recipient eggs are transferred to the operating dish.

Concerning terminology: Unfertilized eggs of course are not "enucleated." The first polar body is given off while the egg is in the oviduct and the egg is in metaphase of the second meiotic division when ova are extruded. Meiosis is resumed with insemination or parthenogenetic activation. Thus, there is no nucleus and what is "enucleated" is a spindle, its meiotic chromosomes, and some pigment-containing cytoplasm.

Germinal Vesicle Enucleation—a Special Procedure for a Giant Nucleus. The germinal vesicle is the enormous nucleus of ovarian eggs. It is so large in some amphibian eggs that it can be seen with the unaided eye after dissection. It cannot be removed from an oocyte with either the Porter manual method or the laser irradiation method described above. Ovarian eggs are not normally the recipients of implanted nuclei in conventional cloning studies. However, a method for germinal vesicle enucleation is described in a study of oocyte maturation and may be of interest to cloners. The procedure is a modification of a technique devised for the immature eggs of sturgeons and consists of a cut made by a fine knife above the germinal vesicle in the animal hemisphere. The nucleus moves to the surface of the cut and after a few minutes it can be gently nudged out of the oocyte (Dettlaff, Nikitina, and Stroeva, 1964).

Preparation of Donor Cells for Transfer

Dissociation of embryonic fragments or pieces of adult tissue is all the pre-transplantation preparation necessary for donor cells. Dissociation procedures vary with tissue type and several procedures will be described.

The nuclear transplantation technique for *R. pipiens* was developed as a means of appraising the extent to which a nucleus derived from a developing embryo could interact with undifferentiated enucleated cytoplasm. To the present time, a great majority of cloning studies have involved an analysis of nuclei derived from embryos. Accordingly, the dissociation of embryonic cells will be considered first.

Dissociation of Blastula Cells. Blastula and, to some extent, parts of later embryos are easily dissociated in calcium- and magnesium-free electrolyte solutions (e.g., modified Niu-Twitty solution, Appendix III). The use of chelating agents (ethylene diamine tetraacetic acid, EDTA) or proteolytic enzymes is not necessary for complete dissociation of blastula cells and some cells from later embryos. Dissociation of blastulae is accomplished by removing the jelly manually (described previously), removing vitelline membrane, and dissecting the animal hemisphere from the yolky vegetal hemisphere cells with glass microneedles. It is conventional procedure to utilize only the animal hemisphere cells since many of them are of diameter appropriate for cloning. The cells become spherical after dissociation and many sizes of cells are available for transfer from an animal hemisphere. Choice of an appropriate cell size is discussed below. A small Stender dish filled with calcium- and magnesium-free electrolyte solution is appropriate for dissociation. Dissociation of all animal hemisphere cells except the cortical (superficial) layer occurs within 20 to 30 minutes at 18° to 20°C.

Dissociation of Cells from Gastrulae and Later Embryos. Dissociation procedures for *Rana* and *Xenopus* embryos beyond the blastula stage do not vary substantially from those described for blastulae. Gastrulae disaggregate in calcium- and magnesium-free solution but the presence of a chelating agent (EDTA) and pH 7.8 to 8.0 are required in order to obtain free cells from post-gastrula embryos up to the neural tube stage (Jones and Elsdale, 1963).

Anuran (*R. pipiens*) and a variety of urodele neurulae dissociate by treating cultures with KOH, added a drop at a time until pH 9.6 to 9.8 is attained (Townes and Holtfreter, 1955). More recently, good disaggregation has been obtained by treating tissue fragments of embryos in a calcium-free Holtfreter's solution (Appendix III) to which has been added 0.005 M sodium citrate. Trypsin is added (0.2 to 2%) to the citrated calcium-free Holtfreter's solution to disaggregate older embryos (Jacobson, 1967).

Cytochalasin B (CB) causes the dissociation of cells in the embryos of *Xenopus* (Bluemink, 1971) and *R. pipiens* (Schaeffer, Schaeffer, and Brick, 1973). Gastrula cells are reported to round up and retract from each other producing a mass of completely dissociated single cells. Cytochalasin B produces this effect at concentrations ranging from 1 to 10 μg/ml. Dissociation by CB is not thought to be toxic because cells reaggregate spontaneously when the drug is removed. No one to my knowledge has used CB to effect dissociation for nuclear transplantation purposes but it might prove to be useful because the manner of dissociation of donor cells does indeed affect nuclear transplantation results (see discussion in Chapter 3).

Dissociation of Adult and in Vitro Cultured Cells. Nuclei from normal adult cells have been only rarely transplanted. One method of disaggregation involves a preparation of single cells from normal adult frog kidney directly (i.e., without in vitro culture) by disrupting the tissue in a Teflon[R] homogenizing tube with 0.5% trypsin in modified (calcium- and magnesium-free) Steinberg's solution (Appendix III). The cell mixture is centrifuged, rinsed, recentrifuged, and finally resuspended in Steinberg's solution (King and DiBerardino, 1965).

Cultured cells from *Xenopus* embryos and adults have been transplanted. Technically, the cultured cells are not disaggregated. However, release of the cells from their substrate in preparation for cloning is not in its end result different from disaggregation. Donor cells from monolayered cultures were released from their substrate by incubation in a trypsin-saline solution at pH 7.8. The cell suspension was then washed in the saline solution (without trypsin) and transplanted in the same solution (Gurdon and Laskey, 1970b).

Warm weather primary renal tumors of sexually mature adult *R. pipiens* (Chapter 4) and tumor cells from anterior eye implants (DiBerardino, King, and McKinnell, 1963) become loose and then individual cells fall away from the tumor mass within 30 minutes at 18°C in a modified Niu-Twitty or calcium- and magnesium-free Steinberg's solution (Appendix III). These tumor cells are suitable for immediate nuclear transfer without culture.

Nuclear Transplantation

Earlier sections of this account of cloning have described the preparation of microneedles, micropipettes, recipient egg cytoplasm, and cells to serve as nuclear donors. It will be the purpose of this section to describe the bringing together of the previously prepared cell parts with the microinstruments in the procedure known as cloning.

Concerning the Nuclear Transfer Solution. No "normal" nuclear medium is yet available (Briggs and King, 1953). This unhappy comment on the state of cloning technology is supported by experiments on oocyte nuclei (germinal vesicles). A method for germinal vesicle isolation is described with an account of their chromosomes and nucleoli in Duryee (1950). Germinal vesicles swell when isolated in saline solutions of various osmotic and ionic compositions (Battin, 1959; Hunter and Hunter, 1961). Damage other than swelling occurs with isolation. Nucleoli are dissolved when germinal vesicles are cleaned with distilled water, saline, or sucrose solutions. Nucleoli are lost in less than a minute when the nuclear isolation medium lacks calcium or if sodium oxalate is added to a calcium-containing saline solution (Holtfreter, 1954). An "explosive" collapse of nuclear gel structure has been described if the medium contains calcium salts (Callan, 1952). If these studies are relevant at all to cloning technology, it would seem that you lose with or without calcium. However, a way of contriving an adequate nuclear medium has been suggested. Reversible enucleation, i.e., removal of a nucleus for exposure to a test solution, followed by reintroduction into the test cell to see if the reconstituted cell behaves properly, has been explored with amoebae (Burnstock and Philpot, 1959). Regrettably, these studies concern nuclei of amoebae, which are exceptionally resistant to damage during isolation. So far, no extensive investigation has been made concerning the effects of physiological media on embryonic nuclei for cloning purposes.

Since naked nuclei swell and are damaged in other ways when placed in certain saline solutions, it becomes crucially important that great care be used in selecting a minimally damaging transfer solution and that vigilance be exercised to protect the transferred nucleus with its own minimally disturbed cytoplasm (lacking a protective nuclear medium, the cloner must rely upon the shielding effect of the cell's own cytoplasm). Eight media were tested with respect to their effect on the cleavage and later developmental potentialities of nuclei transplanted to enucleated eggs. Included were the solutions of Niu and Twitty, Holtfreter, Duryee, Kassel, Wilde, and others. The highest yield of complete embryos was obtained when Niu-Twitty solution (Appendix III) was used (Briggs and King, 1953). Legname and Barbieri (1968) obtained about twice as many blastulae by nuclear transfer with Barth and Barth's solution (Appendix III) as they did with Niu-Twitty solution. However, the percentage of experimental embryos that developed to the tadpole stage was about equal with both solutions. These results would seem to indicate that while Barth and Barth's and Niu-Twitty are equally gentle in their effect on nuclei, one (Barth and

Barth's) seems to have a sparing effect on the integrity of the extra-nuclear cleavage center. In my laboratory during the next decade, we have been unable to detect any difference between Niu-Twitty solution and Steinberg's solution when used as nuclear transfer media.

Operating Dishes. The dissociated cells and prepared eggs are placed in a common operation dish. One kind of operation dish is a Syracuse dish containing a mixture of waxes and lampblack. One or more depressions of appropriate size are made to hold the recipient activated and enucleated egg. Because dissociated embryonic cells tend to stick to the wax bottom and because the delicate tip of a micropipette breaks on touching the firm wax, a blastula animal hemisphere is dissected and placed pole surface up in the dish. The dissected animal hemisphere serves as a platform for dissociated donor cells. Embryonic cells do not stick to the dark cortical surface of the blastula platform, which also provides good optical contrast to the light-colored, dissociated embryonic cells. The platform provides a resilient surface that does not break the micropipette on contact. This dish is similar to the operating container described by King (1966, 1967).

A possible complication associated with blastula platforms in wax-bottomed operating dishes is the hazard of a platform cell becoming loose and intermingling with a group of cells to be transplanted. Thus, it is possible that the operator may believe he is transplanting exclusively cells derived from a nucleocytoplasmic hybrid blastula when in reality his transplantations may include one or more normal cells that floated loose from the blastula platform.

Problems concerning wax-filled dishes may be obviated with operating dishes filled with 2% agar made up in the transplantation medium (Niu-Twitty or Steinberg's, Appendix III) and darkened with decolorizing charcoal (e.g., Norite). The agar is mixed with the saline solution and charcoal and brought to a boil. It is poured into 60 mm diameter Petri dishes to a depth of about 4 mm. The agar is allowed to gel; the dishes are covered and stored in a refrigerator at 4°C until needed. A week's supply of agar operating dishes can be made in a few minutes. Blastula and later embryonic disaggregated cells do not stick to the agar; excessive light is not reflected into the operator's eyes because of the charcoal; and the delicate tips of micropipettes seem not to be damaged in the soft agar substrate.

Because of the extremely small size of adult cells with the possibility of their becoming lost in a large operating dish, it is more convenient to keep them in a separate container from the recipient eggs. The individual adult cell is picked up in the micropipette; the micromanipulator is racked up to clear the donor cell container; the operating dish containing the darkened 2% agar and recipient eggs covered with transfer solution is rapidly placed under view of the dissecting microscope; the micromanipulator is racked back into position; and transfer of the nucleus is effected in the customary manner. The container for adult donor cells may be as simple as a large drop of solution on a glass microscope slide. A small Stender dish or Petri dish may also prove to be suitable. Lighting from below (i.e., transmitted light) may be necessary to see clearly adult cells, thus necessitating a clear glass or plastic container.

Hardware for Cloning. The nuclear transplantation apparatus consists of a binocular stereoscopic microscope and a micromanipulator that is adapted to hold a micropipette and syringe assembly (Figure A-29). The microscope field of view is lighted with two lamps (e.g., American Optical Universal microscope lamps or equivalent lamps) and the microscope is focused with foot controls, which allows the operator free use of his hands to manipulate donor cells and recipient eggs. Focusing is either mechanical (King, 1966, 1967) or by an electrical motor (Gomco Electra-Focus[7]).

A variety of micromanipulators are available and suitable for nuclear transplantation (Kopac, 1964). Instruments manufactured by Emerson, Brinkmann, Leitz, and other companies are illustrated and described by El-Badry (1963). The Singer microdissector as used in *Xenopus* nuclear transplantation experiments is illustrated in Elsdale, Gurdon, and Fischberg (1960). An Emerson micromanipulator[8] modified to allow for increased mobility of the micropipette holder in the vertical plane is used in my laboratory.

I use a Gordon syringe and micropipette holder,[9] which is an exceedingly rugged microinjection apparatus (Gordon and Malacinski, 1970). The syringe and part of the connecting tube are filled with water. There is an air cushion between the fluid in the distal

Figure A-29. Nuclear transplantation apparatus for *R. pipiens*. The dissecting microscope is focused by means of a foot-actuated electric motor, which frees the operator's hands to operate other controls. The operating stage is cooled by an electric cold plate. The micropipette is positioned by a micromanipulator and the meniscus within the micropipette is moved by means of the machined syringe. (From McKinnell, 1973b.)

end of the micropipette and the water in the part of the tubing that connects with the syringe assembly. Thus there is less compressible air in the system than in a syringe filled entirely with air. An experienced operator has no trouble moving the meniscus in the micropipette a few micrometers at a time with the Gordon apparatus. A different injection system is described in the section on *Xenopus* cloning later in this appendix.

Cool temperature of the laboratory and solutions is important for good nuclear transplantation results (Hennen, 1970). Because of this, I use an air-conditioned laboratory kept between 11° and 14°C and as an added precaution for keeping the *R. pipiens* eggs and disaggregated cells cool, I use a refrigerated operating stage.[10]

Micromanipulation of Donor Cell. The description that follows will apply to the transplantation of blastula nuclei of *R. pipiens*. Appropriate modifications may be made to adapt the procedure to the transplantation of nuclei of other cell types.

Activated enucleated eggs and a dissociated blastula are prepared as described above. One egg is placed in a depression in the operating dish along with 20 to 50 dissociated donor cells. Reaggregation occurs in a short period of time if too many dissociated cells are placed in an operating dish containing a transplantation medium with calcium and magnesium ions, such as Niu-Twitty or Steinberg's solution (Appendix III).

The micropipette is put into approximate position under view of the microscope with the micropipette holder forming a 45° angle with the horizontal plane. Transplantation fluid from the dish is drawn up into the micropipette forming a meniscus about 3 mm from the micropipette tip. The tip is maneuvered until it just touches a cell about ¼ larger in diameter than the diameter of the micropipette. A cell of this size is chosen because the plasma membrane is ruptured with the cell is drawn into the micropipette (Figure A-30). Rupturing is important because a transplanted cell with an intact plasma membrane is developmentally isolated from the host cytoplasm (Briggs and King, 1953; King and McKinnell, 1960). Some practice is required in picking up and rupturing cells for transfer. Excessive dispersal of shielding donor cell cytoplasm results if the cell is drawn into the micropipette too rapidly. If the cell to be transplanted is drawn into the micropipette too slowly, however, it tends to deform to the shape of the micropipette with its plasma membrane remaining intact. Egg cytoplasm and donor cell cytoplasm have a firm consistency at low temperature. It is therefore easier to rupture the plasma membrane when picking up the cell when the laboratory is cooled. Ideally, the nucleus is liberated by plasma membrane rupture but protected by an absolute minimal cytoplasmic dispersal.

Nuclear Insertion. The micropipette with its contained nucleus and protective cytoplasm is then placed into position near the surface of the recipient enucleated egg. The egg is situated in a depression that has been formed in the agar or wax substrate. The depression prevents the egg from moving when the micropipette

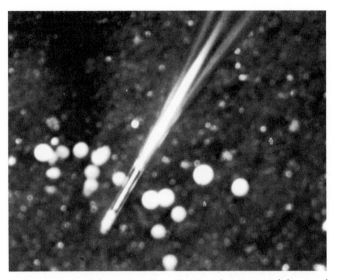

Figure A-30. Dissociated blastula cells and micropipette containing one broken blastula cell. Dissociated cells may be seen to vary in diameter. The cell for transplantation was chosen because it was slightly larger than the orifice of the micropipette. The meniscus of transplantation medium is located above the level of the cell, which is at the tip of the micropipette.

enters the egg. There are alternative ways of introducing the micropipette into the recipient egg but a convenient and simple means is to place the micropipette into position, hold it stationary, and thrust the egg onto the micropipette by moving the operating dish laterally toward the micropipette. The movement of the dish is of course only a fraction of a mm. A relatively rapid movement of the operating dish is required because the surface of the enucleated egg is sufficiently elastic to stretch extensively. The rapid movement results in the sharpened micropipette tip cutting through the surface of the egg with minimal distortion of the egg cortex. The operator should take extreme care to preserve the integrity of the recipient egg cortex. Recent studies have shown that the superficial layer of the amphibian egg is extremely important for normal development—alteration of the cortical layer profoundly affects ontogeny (Brachet, 1972; Brachet and Hubert, 1972; Curtis, 1965; Grant and Wacaster, 1972).

The nucleus and its protective cytoplasm is now hidden in the interior of the recipient egg (Figure A-31). Proper injection is dependent upon visualization of the transplantation fluid meniscus. Sufficient pressure is exerted by the syringe to push the nucleus into the enucleated egg. The amount of pressure necessary is determined by observing the meniscus. The nucleus can be considered to be in place in the interior of the egg when the meniscus has moved down the micropipette a distance equal to the space occupied by the donor cell.

The micropipette is withdrawn from the egg by racking the micromanipulator vertically to raise the egg a few millimeters above the agar or wax substrate. The points of jewelers' forceps are placed one on either side of the micropipette and the egg is re-

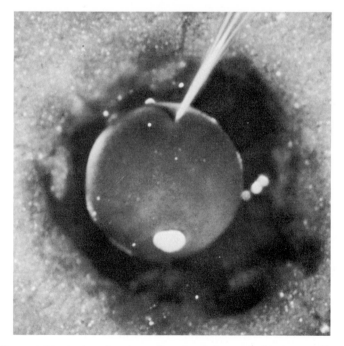

Figure A-31. Activated and enucleated recipient egg with donor cell-containing micropipette thrust into it. The meniscus of the transplantation fluid has been pushed to just outside the surface of the egg, which liberates the broken cell within the egg.

moved by pressing the forceps against the egg and pushing it off of the micropipette.

A channel is sometimes produced that leads from the interior of the egg through the vitelline membrane and residual jelly. Substantial cytoplasmic leakage may occur through the puncture made by the micropipette. It may be reduced by either severing the channel with microneedles as described by King (1966, 1967) or by removing the impaled egg from the micropipette briskly so that yolk has less opportunity to extend out into the jelly to keep the channel open. The operated egg is then transferred with a large mouth dropping pipette to its individual Stender dish filled with 10% magnesium-fortified Holtfreter's solution (Appendix III) and retured from the cold operating room to the 18° laboratory.

The transplanted egg is observed at the time that the first cleavage division should occur. The expected time of cleavage is predicted not from the interval following transplantation but from the time of parthenogenetic activation (pricking with a clean glass needle—described at the beginning of Part B of this appendix). The two-blastomere stage is expected in about 3½ hours following activation when the operated egg has been kept at 18° following nuclear insertion (Shumway, 1940, Appendix VI). The time of first cleavage is recorded because it is not unusual for cleavage to be delayed for about 1 hour. Polyploidy is almost invariably associated with delays in cleavage (Briggs and King, 1952; Sambuichi, 1959; McKinnell, 1964).

Nuclear transplant animals that ensue from experiments with early embryonic nuclei can frequently be reared to metamorphosis. Rearing methods are described in detail in Appendix I.

Nuclear Transplantation in Other
Species of Amphibia

Xenopus. The procedure of Briggs and King (1952) developed for *R. pipiens* was modified for the South African clawed frog (also known as the platanna), *Xenopus laevis* (Figure A-32). Usefulness of *X. laevis* in the laboratory is attested to by the fact that eggs are available throughout the year (they are available from mid-October to the following spring in *R. pipiens*) and that sexual ma-

Figure A-32. The South African clawed frog *Xenopus laevis*. Photograph by Gordon A. F. Dunn (from Nace et al., 1974).

turity is reached at 12 months of age. Commercial sources, rearing methods, and other notes concerning husbandry of South African clawed frogs are found in Gurdon (1967b), Deuchar (1972, 1975), and Nace et al. (1974).

Cloning in *X. laevis* was reported by Fischberg, Gurdon, and Elsdale (1958a). The technique with *Xenopus* differed in the earliest experiments from that utilized in *Rana* studies in that the host egg was not activated and enucleated. The authors believed that enucleation was not necessary because their data indicated that the maternal pronucleus participated in development in only about 1% of the nuclear transfers. In one series of 70 tadpoles produced by nuclear transfer, 79% (55 tadpoles) were diploid and 17% (12) tadpoles were tetraploid. The remaining 4% were mosaic. No triploid embryos were recorded (as would be expected if the maternal haploid pronucleus were to fuse with the introduced nucleus) nor were any haploid embryos reported. Haploidy would result if the transplanted nucleus failed to develop and if the maternal pronucleus were stimulated to develop parthenogenetically. Accordingly, it would seem that enucleation would not be necessary, especially if the yield of successful nuclear transplant embryos remained substantially above 1%. In the early experiments with *X. laevis*, the yield of nuclear transplant embryos was indeed in excess of 1%. Eighty-nine embryos (10%) hatched normally and 61 embryos (7%) started feeding from among 905 attempted nuclear transplantations.

However, within a short period of time, the usefulness of egg pronuclear inactivation with ultraviolet irradiation was recognized (Gurdon, 1960a). The method of enucleation of the recipient egg remains a major technical difference in the methodologies developed for the two species of anurans. There have been recent reports questioning the reliability of enucleation by ultraviolet irradiation (DuPasquier and Wabl, 1977; McAvoy, Dixon, and Marshall, 1975). No cytoplasm is lost with ultraviolet inactivation of the egg chromosomes and the irradiation is claimed to cause no permanent damage to the egg cytoplasm.

Dissociation of donor embryo cells is accomplished in a calcium- and magnesium-free solution (Barth and Barth's solution X, Appendix III) containing versene.

The apparatus employed in *Xenopus* cloning studies is not dissimilar to that used by many experimenters who utilize *Rana*, i.e., it consists of a binocular dissecting microscope, micromanipulator, and microsyringe (illustrated in Elsdale, Gurdon, and Fischberg, 1960). The micromanipulator is an Oxford manipulator (also known as a Singer Microdissector)[11] which is not at all different in principle from any of those micromanipulators used with *Rana*. The Oxford instrument simply reduces all hand movements by a ration of 5:1 and does not reverse movement in any direction.

The injection microsyringe assembly does, however, vary somewhat from those commonly used in North America with the leopard frog. The Agla micrometer syringe[12] and the 1 mm bore plastic tubing that connects to the micropipette are filled with non-toxic paraffin oil. The oil is of course non-compressible. The oil extends out into the micropipette. At the time of transfer, a minute bubble of air is drawn into the micropipette followed by the transplantation fluid. The bubble of air serves as a marker so that the operator can observe its movement as he inserts the donor cell into the recipient egg (much as the operator using *Rana* observes the meniscus in his transplantation micropipette at the time of nuclear insertion). The micrometer syringe, semi-rigid tubing, and non-compressible paraffin oil have the advantage of positive control at the time of donor cell pickup. It may be of interest to note that the same combination of Oxford manipulator and Agla syringe is currently being used by mammalian experimental embryologists for extrauterine transplantation of embryos (see Kirby, 1971, figure 9-6).

Much of the above discussion pertains only to the cloning of embryonic nuclei. Lighting conditions, microscopy, solutions, and other technical considerations are described for the transplantation of cultured embryonic and adult *Xenopus* cells by Gurdon and Laskey (1970b).

Another technical difference between the two procedures involves the environment of the egg at the time of nuclear insertion. Nuclear transfer in *Rana pipiens* is accomplished with the recipient egg covered with an electrolyte solution. *Xenopus* procedure differs in that the eggs are not surrounded by a fluid when they receive their implanted nuclei.

That development of the recipient egg is controlled by the transplanted nucleus and not the egg nucleus is now substantiated by the use of a mutation in *Xenopus* which suppresses nucleolus formation. Individuals that are homozygous for the nucleolar mutation have cells with no normal nucleoli (they possess only small droplets of nucleolar material) and the tadpoles die before feeding. Heterozygous embryos undergo normal development but are distinguished from the wild-type by the presence of only one nucleolus per cell instead of the normal two. Development of the transplanted nucleus can be verified by utilizing donor nuclei obtained from heterozygous cells transplanted to irradiated eggs of wild-type female frogs. The presence of one nucleolus per cell (which can quickly be detected with phase contrast microscopy) is evidence that the transplanted nucleus is the developmentally active nucleus. A description of this exceptionally useful *Xenopus* mutation is found in Elsdale, Fischberg, and Smith (1958) and in Hay and Gurdon (1967).

As with *Rana*, embryos produced by cloning in *Xenopus* often develop abnormally. Experiments were designed to reveal whether the abnormalities are due to innate qualities of donor nuclei or whether they may be ascribed to technical factors (Gurdon, 1960b). Three kinds of dissociation solutions were studied—each gave similar results with respect to abnormalities among the resulting embryos. It was noted, however, that thorough dissociation was essential to minimize abnormalities. The extent of distortion of a donor cell in the micropipette affects the yield of normal embryos. Surprisingly, too little distortion results in a low yield. Little distortion presumably results in incomplete plasma membrane disruption. Excessive deformation results in exposure of the nucleus to the electrolyte transplantation medium. Optimal shape change of the donor cell in the micropipette is just that required to break the cell membrane.

The volume of electrolyte solution injected into the recipient egg with the donor cell may be increased by 2 or 3 times the volume of the transplanted cell with apparently no deleterious effects on the resulting embryo.

Xenopus eggs have substantial innate variation. Not only is there variation in egg quality between different females but the quality of eggs from an individual female may vary with time. *Xenopus* obtained from the wild do not necessarily provide better eggs than South African clawed frogs bred in the laboratory.

Urodeles. The tailed amphibian with the greatest number of known genetic variants is the Mexican axolotl, *Ambystoma mexicanum* (Figure 5-8) (Briggs, 1973; Malacinski and Brothers, 1974). The Mexican axolotl is a neotonous salamander that is aquatic and easily maintained in the laboratory (Fankhauser, 1967; Nace et al., 1974). Axolotls reach sexual maturity in 10 to 12 months and a mature female may be bred every 3 months. Several hundred eggs are obtained at each spawning and the eggs, because of their enormous size (ca. 2 mm diameter), are manipulated with ease. The genetic traits and simple husbandry of *A. mexicanum* led to the development of a nuclear transplantation procedure (Signoret, Briggs, and Humphrey, 1962). A similar method was devised for the Triton, *Pleurodeles waltlii* (Signoret and Picheral, 1962).

Unmated female axolotls injected with pituitary gonadotropin will produce eggs. Spawning occurs at the rate of 30 to 60 eggs per minute 20 to 24 hours after injection. Ova are decapsulated manually and activated either by heat shock (5 minutes at 35°C) or by electric shock (Signoret and Fagnier, 1962; Briggs, Signoret, and Humphrey, 1964). In early experiments, urodele eggs were activated with irradiated sperm (Lehman, 1955). Before activation the egg pigments are unevenly distributed. Within 1 or 2 hours following activation the pigments become more uniformly dispersed and the second polar body appears. Enucleation is accomplished by ultraviolet irradiation of animal hemispheres for 4 or 5 minutes after activation. That this is an effective procedure for the elimination of maternal chromosomes is proved by the irradiation of inseminated ova, all develop as androgenetic haploids. The fate of ultraviolet irradiated maternal genetic material that does not participate in development is reported by Bideau (1964). Alternatively, the egg nucleus may be sucked out through a puncture in the egg membrane (Curry, 1936; Lehman, 1955).

The actual nuclear insertion procedure does not vary significantly from that described for *R. pipiens*. However, following the insertion of a donor cell with undispersed cytoplasm in an activated and irradiated egg, there is considerable leakage for several hours from the puncture left in the vitelline membrane by the micropipette. A dumbbell-shaped glass plug (25 μ diameter and 0.8 mm long) inserted into the micropipette puncture effectively stops leakage and appears not to interfere with development.

That development of transplanted eggs is due to the inserted embryonic nucleus and not due to maternal genes was verified by a black pigment pattern in all of 37 experimentally produced larvae resulting from nuclei carrying a dominant gene for black pigment combined with recipient eggs from homozygous recessive white females. The transplantation of genetically marked nuclei in the axolotl (Briggs, Signoret, and Humphrey, 1964) is analogous to the use of nuclear markers in anurans (McKinnell, 1960; Simpson and McKinnell, 1964; Elsdale, Gurdon, and Fischberg, 1960).

A film "Nuclear Transplantation" produced by C. M. Flaten and A. J. Brothers with a copyright date of 1976 is available from Indiana University (16 mm / 12 minutes / color / order number NSC 1448). The film utilizes photomicrography and time lapse sequences to reveal the nuclear transfer procedure for the Mexican axolotl.

Notes on Teaching Methods

The transplantation procedure for amphibian nuclei was devised by Briggs and King at The Institute for Cancer Research in Philadelphia. Some investigators made pilgrimages to that place (or to other laboratories where cloning is practiced) to learn the technique. Some were self-taught. It is my opinion that there is an economy of effort if a student works for a short period of time under the guidance of an experienced nuclear transplanter. Thus, many pitfalls may be averted by forewarnings from the journeyman.

I have found one teaching device to be exceptionally useful for the craft. While it is simple to show the apprentice methods for storage of frogs, induction of ovulation, fertilization, microimplement manufacture, and other operations associated with the technique, it is more difficult to demonstrate the actual preparation of host eggs and the transfer to them of nuclei. I have found that an American Optical Cycloptic dual observation microscope[13] (Figure A-33), which provides simultaneous stereoscopic vision with final common optical path for both workers, is advantageous for the purpose of nuclear transfer demonstration. Light from the object being viewed is split so that both observers see identical images. This results in less illumination than a conventional dissecting microscope but the light diminution becomes apparent only at high magnifications.

An alternative microscope for teaching purposes is one that is manufactured by Leitz and described by Freund (1966). This consists of two Greenough-type microscopes with a common mounting. Lighting with this system is superior to that of the split optical path dual microscope but the usefulness of the microscope suffers from the fact that teacher and student have slightly different views of the material under observation.

Only one or two teaching sessions are required to acquaint the novice with the enucleation technique and proper methods for nuclear transfer. The student can then develop skill with solo prac-

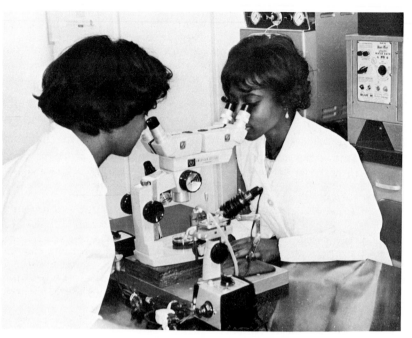

Figure A-33. Technologist experienced in amphibian reproductive biology (*left*) is demonstrating micromanipulative procedure by means of a dual microscope with final common optical path.

tice after the introduction to the technique with the dual microscope.

Low-priced closed circuit television cameras, monitors, and tape decks are now available. I am in the process of developing closed circuit tapes of cloning procedures. When these become available, the student can watch a particular step over and over again and need not intrude on the laboratory chief while doing so. The TV tapes combined with personal instruction with the dual microscope ought to reduce the time needed to learn cloning.

Notes

1. The Livingston micropipette puller is available from F. Stanley Hockman Scientific Instruments, 453 South New Middletown Road, Media, Pennsylvania 19063.

2. Precision drawn soft glass rod and capillary tubing are available from Drummond Scientific Company, 500 Parkway, Broomall, Pennsylvania 19008.

3. The manufacture of relatively large micropipettes suitable for amphibian nuclear transfer is described in this section. Readers should be alerted to the fact that much smaller micropipettes can be fabricated. Beveled micropipettes as fine as 0.1 μ diameter have been described (Brown and Flaming, 1974).

4. A microforge costing substantially less than commercial models can be assembled in any good instrument shop. Specifications and diagrams for an inexpensive microforge have been described by Gurdon (1974a).

5. An alternative method for producing a Pettus slide is to pour some powdered abrasive (Carborundum Powder, 150 grit, available from Fisher Scientific Company, Fair Lawn, New Jersey) onto a clean glass plate. Water is added to form a slurry with the material. The microscope slide is then rubbed in a circular motion with light pressure in the slurry until the slide is frosted.

6. The biolaser system is available from Hadron, 800 Shames Drive, Westbury, New York 11590. It is a simple task to assemble an inexpensive microbeam Helium-Neon gas laser with continuous radiation in the red at 6,328 Å (Lacalli and Acton, 1972). It is my judgment that this apparatus would serve in lieu of the more expensive ruby pulse laser available commercially. However, its efficacy in enucleation of frog eggs has not been tested.

7. This is available from Gomco Surgical Manufacturing Corporation, 828 East Ferry Street, Buffalo, New York 14211.

8. The Emerson micromanipulator is available from J. H. Emerson Company, 22 Cottage Park Avenue, Cambridge, Massachusetts 02140.

9. The holder is available from Gordon Machine Shop, 1908 Viva Drive, Bloomington, Indiana 47401.

10. The TCP-2 Thermoelectric Cold Plate is available from Thermoelectrics Unlimited, Inc., 1202 Harrison Avenue, Holly Oak Terrace, Wilmington, Delaware 19809.

11. The Oxford manipulator is available from Micro Techniques (Oxford) Ltd., 7 Little Clarendon Street, Oxford OX1 2HP, England.

12. The Agla micrometer syringe is available from Wellcome Reagents, Ltd., Wellcome Research Laboratories, Beckenham BR3 3BS, England.

13. This is available from American Optical Company, Instrument Division, Buffalo, New York 14215.

Formulae of Solutions Commonly Used in Nuclear Transfer Experiments

(1) Niu-Twitty Solution (Flickinger, 1949)

Solution A	*Solution B*	*Solution C*
500 ml	250 ml	250 ml
NaCl . . . . 3,400 mg	$Na_2HPO_4 \cdot 12H_2O$. 110 mg	$NaHCO_3$ . . . . . . 200 mg
KCl . . . . 50 mg	KH_2PO_4 . . . . 20 mg	
$Ca(NO_3)_2 \cdot 4H_2O$ . . 80 mg		
$MgSO_4 \cdot 7H_2O$ . . . 100 mg		

The three solutions are brought to a boil separately and mixed after cooling.

(2) Modified Niu-Twitty (King, 1966)

Solution A	*Solution B*	*Solution C*
500 ml	250 ml	250 ml
NaCl . . . . 2,943 mg	Na_2HPO_4 . . . . 1,300 mg	$NaHCO_3$ . . . . . . 200 mg
KCl . . . . 50 mg	KH_2PO_4 . . . . . . 116 mg	

This solution contains phosphate buffer but lacks calcium and magnesium ions and is used for dissociation of embryonic cells. The three solutions are brought to a boil separately and mixed after cooling.

(3) Amphibian Ringer's Solution (Rugh, 1962)

NaCl . . . . 6,600 mg
KCl . . . . 150 mg
$CaCl_2$ . . . . 150 mg

NaHCO$_3$ . . . . 100 mg
H$_2$O to make . . . . . . 1,000 ml

A white insoluble precipitate forms if the solution is autoclaved.

(4) Holtfreter's Solution (Holtfreter, 1931)

NaCl . . . . 3,500 mg
KCl . . . . 50 mg
CaCl$_2$ . . . . 100 mg
NaHCO$_3$ . . . . 200 mg
H$_2$O to make . . . . . . 1,000 ml

(5) Magnesium-Fortified Holtfreter's Solution (Brown and Casten, 1962)

NaCl . . . . 350 mg
KCl . . . . 5 mg
CaCl$_2$ . . . . 10 mg
MgCl$_2$·6H$_2$O . . . . 20 mg
H$_2$O to make . . . . . . 1,000 ml

This solution is 1/10 Holtfreter's solution plus magnesium, which is required for tadpole growth.

(6) Matthaei-Nirenberg Medium (Briggs, Signoret, and Humphrey, 1964)

KCl . . . . 4,473 mg
Mg(C$_2$H$_3$O$_2$)$_2$·4H$_2$O . . 2,145 mg
Mercaptoethanol . . . . . 470 mg
Tris buffer . . . . 1,211 mg
H$_2$O to make . . . . . . 1,000 ml

(7) Moore's Modification of Barth and Barth Solution (Moore, 1960a)

Stock A

NaCl . . . . 5,150 mg
KCl . . . . 75 mg
MgSO$_4$·7H$_2$O . . . . 204 mg
Ca(NO$_3$)$_2$·4H$_2$O . . . . . . 78 mg
CaCl$_2$·2H$_2$O . . . . 60 mg
H$_2$O . . . . 600 ml

Stock B

NaHCO$_3$ . . . . 400 mg
H$_2$O . . . . 500 ml

Stock C

Na$_2$HPO$_4$ . . . . 60 mg
KH$_2$PO$_4$ . . . . 75 mg
H$_2$O . . . . 500 ml

Stock V

NaCl . . . . 5,150 mg
KCl . . . . 75 mg
Versene* . . . . 170 mg
H$_2$O . . . . 600 ml
*Disodium ethylenediamine tetraacetate

The operating solution is prepared as follows: 60 ml of Stock A, 25 ml of Stock B, and 25 ml of Stock C are put in separate flasks, autoclaved (20 minutes at 15 pounds pressure), cooled, and then combined. The dissociating solution is prepared as follows: 60 ml of Stock V, 25 ml of Stock B, and 25 ml of Stock C are put in separate flasks, autoclaved, allowed to cool, and then combined.

(8) Steinberg Medium (Steinberg, 1957)

17% NaCl . . . . 20 ml
0.5% KCl . . . . 10 ml
0.8% Ca(NO$_3$)$_2$·4H$_2$O . . . 10 ml
2.05% MgSO$_4$·7H$_2$O . . . . 10 ml
1.0 N HCl . . . . 4 ml
Tris (hydroxymethyl)
    aminomethane . . . . . 560 mg
H$_2$O . . . . 946 ml

Stock solutions of the salts are made up in advance to facilitate rapid preparation of the medium. The solution is then made up in one flask and autoclaved, there being no pH change in the process.

(9) Modified Steinberg Medium (King, 1966)

17% NaCl . . . . 20 ml
0.5% KCl . . . . 10 ml
1.0 N HCl . . . . 4 ml
Tris (hydroxymethyl)
    aminomethane . . . . . 691 mg
H$_2$O . . . . 966 ml

Modified Steinberg's medium is a calcium- and magnesium-free medium for the dissociation of early embryo cells.

(10) Barth and Barth Solution (Barth and Barth, 1959)

Solution X

A

NaCl (Merck Biol.)	5.150 g
KCl	0.075 g
$MgSO_4 \cdot 7H_2O$	0.204 g
$Ca(NO_3)_2 \cdot H_2O$	0.062 g
$CaCl_2 \cdot 2H_2O$	0.060 g
H_2O to make	500 ml

B

$NaHCO_3$	0.200 g
H_2O to make	250 ml

C

Na_2HPO_4	0.300 g
KH_2PO_4	0.375 g
H_2O to make	250 ml

Use double distilled water prepared in a glass still during the second distillation. Keep stock solutions refrigerated. To prepare Solution X, aliquots of A, B, and C are autoclaved separately. Take 50 ml of A plus 10 ml glass distilled water to correct for evaporation during autoclaving, 25 ml of B, and 25 ml of C and autoclave for 20 minutes at 15 lbs pressure. After autoclaving and before mixing A, B, and C, add 100 mg serum globulin (Bios) to either C or B and bring to a boil several times. Mix A, B, and C.

(11) Barth and Barth Dissociating Medium (Barth and Barth, 1959)

A

NaCl (Merck Biol.) . . .	6,800 mg
KCl	100 mg
Versene*	744 mg
H_2O to make	500 ml

B

$NaHPO_4$ anhy.	100 mg
KH_2PO_4	20 mg
H_2O to make	250 ml

C

NaHCO₃ . . . . 200 mg

H₂O to make . . . . 250 ml

*Disodium ethylenediamine tetraacetate

Autoclave aliquots of A, B, and C separately. Combine 50 ml of A, 25 ml of B, and 25 ml of C. Use double distilled water throughout.

Organisms Used in Nuclear Transfer Experiments

Part 1. *Amphibia*

Ambystoma mexicanum	Signoret, Briggs, and Humphrey, 1962
Bombina orientalis	Ellinger, 1976; Carlson, 1976
Bufo arenarum	Legname and Barbieri, 1968
Bufo bufo asiaticus	Nikitina, 1969
Bufo bufo gargarizans	Tung et al., 1964
Bufo viridis	Stroeva and Nikitina, 1960
Pleurodeles waltlii	Signoret and Picheral, 1962, Picheral, 1962
Pleurodeles poireti	Gallien and Aimar, 1974
Rana arvalis	Stroeva and Nikitina, 1960: Nikitina, 1964
Rana brevipoda	Kawamura and Nishioka, 1963a
Rana catesbeiana	King and Briggs, 1953
Rana esculenta	Nishioka, 1972b
Rana japonica	Sambuichi, 1959; Kawamura and Nishioka, 1963a
Rana nigromaculata	Kawamura and Nishioka, 1963a; Sambuichi, 1964
Rana nigromaculata brevipoda	Sambuichi, 1957a, 1961
Rana nigromaculata nigromaculata	Sambuichi, 1957a, 1961

Rana ornativentris	Kawamura and Nishioka, 1972
Rana palustris	Hennen, 1965, 1974
Rana pipiens	Briggs and King, 1952
R. pipiens, Burnsi dominant mutant	Simpson and McKinnell, 1964
R. pipiens, Kandiyohi dominant mutant	McKinnell, 1960, 1962b, 1964
Rana plancyi chosenica	Nishioka, 1972c
Rana sylvatica	Moore, 1960a
Rana temporaria	Stroeva and Nikitina, 1960; Nikitina, 1964
Triton palmatus	Lehman, 1955
Triton taeniatus	Lopashov, 1945
Triturus alpestris	Lehman, 1955; Pantelouris and Jacob, 1958
Triturus vulgaris	Sládeček and Romanovský, 1967
Xenopus laevis	Fischberg, Gurdon, and Elsdale, 1958a; Gurdon, 1962b
Xenopus laevis laevis	Gurdon, 1961
Xenopus laevis petersi	Ortolani, Fischberg, and Slatkine, 1966
Xenopus laevis victorianus	Gurdon, 1961
Xenopus tropicalis	Gurdon, 1962c

Part 2. *Organisms Other Than Amphibia*

Acetabularia	Hämmerling, 1934
Amoeba	Comandon and de Fonbrune, 1939b; Lorch and Danielli, 1950; Goldstein, 1976
Apis (honeybee)	DuPraw, 1967
Ciona	Tung, et al., 1977
Drosophila	Illmensee, 1973
Physarum polycephalum	Guttes and Guttes, 1963
Stentor	Tartar, 1961
Carassius auratus, Rhodeus sinensis	Tung, et al., 1963

Sources of Amphibians
for Research. II

George W. Nace and Jeffrey K. Rosen

The frequent requests for a wide variety of amphibian species and the absence of a clearinghouse for information on sources of amphibians led to the publication in 1971 of a list of suppliers (Nace, Waage, and Richards, 1971). The ink was not dry before that list required amendments. Now, six years later, the renewed press of inquiries leads us again to publish such a list. Several attempts were made to contact each source on the previous list, and only those who reported they were active as of October 1977 appear here. The original list cited 59 sources. This new list cites 64, but of these only 27 were on the original list. The first list named sources for 144 species; this list names 195 species.

These statistics might suggest that although sources change, the supply of amphibians is fairly constant. This, however, would be an incorrect conclusion: there have been many changes in amphibian supply during this six-year period.

Dominating the period was a marked change in the availability of animals in the leopard frog group. In the early 1970s, virtually

Note: The authors are in the Division of Biological Sciences and Center for Human Growth and Development, the University of Michigan, Ann Arbor, Michigan 48109. This is contribution #80 from the Amphibian Facility.

This work was supported by the National Institutes of Health through Grant RR 00572 from the Division of Research Resources.

all leopard frogs used in teaching (13,000,000) and research (2,000,000) (Nace, 1970; Gibbs, Nace, and Emmons, 1971) were captured in the wild. Now, nature no longer can meet the need (Nace, Culley, Emmons, Gibbs, Hutchison, and McKinnell, 1974; Wake et al., 1975; unpublished NIH conference of 1975). Thus, one commercial supplier (Emmons, personal communication) reports that in 1970-71 they processed over 10 *tons* of leopard frogs from one western state alone, whereas in 1974-75, less than 250 *pounds* of these frogs were available (Nace, 1976a). Even commercial frog catchers with 30-40 years of experience come up empty-handed (McKinnell, personal communication), and in a recent study of about 75 known breeding sites in Wisconsin, egg clutches were found at only five (Hine, Les, Hellmith, and Vogt, 1975). This reduction in the number of breeding sites was accompanied by reduced fertility rates for the eggs which were laid. During this same period a population near Ann Arbor, Michigan dropped from 85,000 leopard frogs to *zero* (Nace, 1977).

Such changes in commercially exploitable populations of leopard frogs have been widespread and apply equally to other anura throughout the world. How many species have been affected is unknown. The causes for these population declines seem to be multiple (Cooke, 1972) and although reports from commercial suppliers in the fall of 1977 suggest a stabilization of the situation, they do not signal an improvement. In the face of these circumstances, the laws of supply and demand have forced a rapid increase in commercial prices for frogs ($2.25 per dozen for 3-inch *Rana pipiens* in 1972 from one company compared with $10.00 in 1977 from another), a pressure to shift to new teaching organisms such as the fetal pig (Nace, 1976b), and domestically raised *Xenopus* (Eldridge, 1976), and, apparently, a sharp reduction in the exposure of students to live vertebrates.

A second change during this period has been the acquisition of the major family-enterprise commercial frog supply houses, whose proprietors were in personal contact with the frog catchers, by large suppliers of educational materials. This has been accompanied by reduced contact with the frog catchers. Even had the supply of frogs in nature remained unchanged, this new structure of the industry would have reduced the accessible supply of animals. Although many of the individuals have changed, a second category of suppliers remains constant. These are the investigators and "hobbyists" who are willing to share their resources with others. Among these, some are able to provide limited services gratis, whereas others, because of the demand and for institutional reasons, must charge for their services.

A third change during this period is evident in the research use of amphibians. Nace (1968, 1970, 1976a, 1976c) and Gibbs, et al., (1971) called for improvement in the quality of the amphibians used in research. This improvement was badly needed in the areas of systematic definition, disease, nutritional and genetic control, and in-house care of the animals (see Symposium, 1973). Concurrently, the changing status of many research areas including neurology (Fite, 1976, Llinás and Precht, 1976) and nuclear transplantation (McKinnell, 1978) has occasioned an increasing demand for amphibian models with genetic definition and genetic markers.

Important events in the improvement of amphibian standards have included work on the systematic definition of the popularly used leopard frogs. Because the reevaluation of the species in this complex by Pace (1974) is of great utility and importance to the potential users of this list of sources, her description and key to these species is reproduced here:

Rana pipiens Schreber
> This is probably the most widely ranging member of the species complex. It is the only leopard frog known in Canada. In the United States it is found in New England, in New York and Pennsylvania, in most of Ohio and northern Kentucky, in northern and central Indiana, in northern Illinois, in Iowa and west to and through the Rocky Mountains. In the Far West it is found in the Snake River Valley, the Columbia River Valley, and in Lake Tahoe and other areas of California and Nevada. It is also found at higher elevations at least as far south as Alpine, Arizona, and may occur in at least the northern part of Mexico as well.

Rana utricularia Harlan
> Found throughout most of the Atlantic Coastal Plain from southern New York and northern New Jersey and through the Gulf Coastal Plain to somewhere between Corpus Christi and Victoria, Texas. It is found in southern Missouri, southern Illinois, southwestern Indiana, most of Kentucky, and extreme southern Ohio. It is found in southeastern Kansas and

eastern Oklahoma and is the only leopard frog known from Tennessee, Arkansas, Louisiana, Mississippi, Alabama, Georgia, Florida, South Carolina, North Carolina, Virginia and Delaware.

Rana berlandieri Baird
    In the United States this species is known only from southern Texas. It is known from as far north as Johnson County, in central Texas, and San Patricio County, along the Gulf Coast. It may also occur west of the Pecos River (known from Ward County, Texas, whose western border lies along the Pecos River).

Rana blairi Mecham, Littlejohn, Oldham, Brown and Brown
    Found in the central plains and prairie regions of the United States, from eastern Colorado, northeastern New Mexico, northern Texas, Oklahoma (except the southeastern third of the state), most of Kansas, part of Nebraska and Iowa, and in northern Missouri, central Illinois, and in scattered localities further east (western Indiana) and south (southern Illinois).

Key to Adult Males in the *Rana pipiens* Complex from the United States (from Pace, 1974)

1.   Skin at angle of jaw overlying internal vocal sac not differentiated in texture or color from surrounding skin (may be somewhat stretched); no distinct white spots on centers of tympana; dorsolateral folds continuous and not displaced, usually wide and low, but discernible to the point where the leg joins the body; dorsal spots usually ringed with light coloration . . . . *Rana pipiens*

1'.   Skin at angle of jaw overlying internal vocal sac differentiated from surrounding skin in some way (e.g., texture or pigmentation); tympanal spots, dorsolateral folds, and dorsal spots variable 2

2(1'). Mullerian ducts absent, or specimen from Florida 3

2'.   Mullerian ducts present and specimen from Texas. . . *Rana berlandieri*

3(2).   External vocal sacs large, spherical, apparently thin-skinned, lying loose at angle of jaw when not inflated or from Florida; dorsolateral folds usually continuous and not displaced . . . . *Rana utricularia*

3'.   External vocal sacs small, usually visible only because skin at angle of jaw is conspicuous when internal vocal sac is not inflated owing to texturings of the skin below the labial stripe; dorsolateral folds usually discontinuous and displaced medially. . . . . *Rana blairi*

It is unfortunate that no key is available for the females of these species. Also, please note that some systematists prefer *R. sphenocephala* to *R. utricularia*. Of additional importance is the recent work on the leopard frogs of the Southwest and Mexico by Frost and Bagnara (1976). However, few, if any, commercial suppliers are prepared to distinguish among any of these species of leopard frogs when they provide *"Rana pipiens."* Indeed, some suppliers attempt to meet demands with frogs imported from Europe, and this may not always be indicated on the packing slips. Thus, the user is urged to exercise caution regarding the identity of amphibians used in research protocols (Bagnara and Frost, 1977).

Another event of importance in defining research amphibians and their use was the appearance of "Amphibians: Guidelines for the breeding, care and management of laboratory animals" (Nace et al., 1974). Although already out of date in some particulars, this document remains the best currently available source of information on the matters suggested by its title. Of particular note, it defines a nomenclature for a classification to permit investigators to select animals most appropriate to their needs, to communicate effectively with suppliers, and to accurately report their data. This nomenclature is reproduced here for the reader's convenience, but the reader is urged to become familiar with the definitions of these terms as given on pages 27-36 of the reference.

1. Wild
2. Wild-caught
   a. Wild-caught nonconditioned
      (1) Wild-caught nonconditioned nontreated
      (2) Wild-caught nonconditioned treated
      (3) Wild-caught nonconditioned miscellaneous
   b. Wild-caught conditioned
      (1) Wild-caught conditioned larvae
      (2) Wild-caught conditioned juveniles or adults
      (3) Wild-caught conditioned miscellaneous
3. Laboratory reared
   a. Laboratory-reared standard
   b. Laboratory-reared miscellaneous
4. Laboratory bred
   a. Laboratory-bred standard
   b. Laboratory-bred miscellaneous

Among laboratory-reared and laboratory-bred animals the following types of populations and lines may be designated:

(1) Random mating lines
(2) Heterozygous isogenic clones
(3) Heterozygous marked lines
(4) Mutant lines
(5) Inbred lines
(6) Gynogenetic diploid lines
(7) Homozygous lines
(8) Haploid animals
(9) Polyploid animals

Attention is also directed to Müller (1976) which on a comparative basis details many experimental parameters of concern to those using amphibians for research and describes the useful series of books on amphibian physiology represented most recently by Lofts (1976).

Among the sources cited in the following list, those marked with "§" are, at present, adhering to the conventions of this classification, and #39 is developing the capacity to do so.

The provision of animals with genetic definition or genetic markers is most advanced at the Axolotl Colony (#3) (Malacinski and Brothers, 1974), as might be expected from the long history of culture of the Mexican axolotl. Attention of the readers is drawn to the informative Axolotl Newsletter available from that resource. Other species for which there has been significant development of genetic definition are *Xenopus laevis* and *Pleurodeles waltlii*. Although at this time no specific sources can be cited for genetically defined stocks of these animals, they are available on occasion from the investigators using them. Among the ranids, the best stock of genetically significant animals is held by the Laboratory for Amphibian Biology, Hiroshima, (#27) and to a lesser extent by the Amphibian Facility.

A major problem in providing high-quality animals to users has been the inadequacy of the care in users' laboratories. Basic to this has been inappropriate housing. The development of housing units that would reduce labor and space costs, promote the health of animals, and be adaptable to a wide variety of species has been a major concern of the Amphibian Facility. Such units are now available and feature a highly adaptable enclosure (amphibians are not held in aquaria, terraria, cages, or pens) called the Michigan

Environmental Enclosure for Small Animals (MEESA) and include an Adjustable Depth Overflow Pipe (ADOP). Figure 1 shows a two-enclosure model made of plastic and Figure 2 shows a six-enclosure unit made of stainless steel. The Amphibian Facility has recently been remodeled and animals are housed in these enclosures mounted on five-tier racks. If these enclosures are used, nearly all aspects of the environment may be controlled. The units are manufactured by Keyco Co. (P.O. Box 12, Peach Bottom, Pennsylvania, 17563, 717-548-2185, Mr. Howard Fricke, President) and are currently marketed by NASCO (#37). We also call the reader's attention to the new enclosures for bullfrogs developed by Culley (1977).

The Amphibian Facility continues to develop and provide lab-

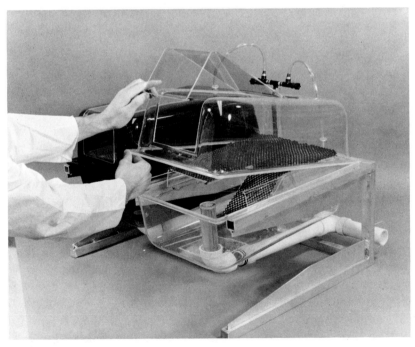

Figure 1. A two-enclosure table top, plastic model of MEESA suitable for carrying up to 150 mature *R. pipiens* for several weeks and 50 indefinitely in each enclosure. Modular units may be added. It can also house all other species of amphibians as well as fish, lizards, snakes, and small mammals or birds. Temperature, water, air flow, and light can be controlled.

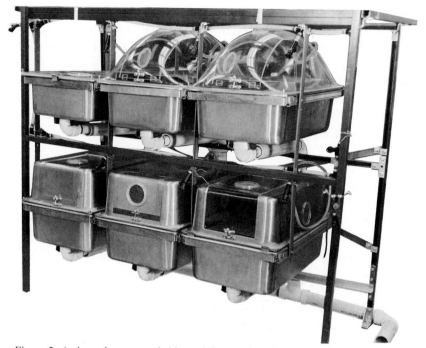

Figure 2. A six-enclosure, stackable, stainless-steel model of MEESA showing several alternative configurations. Stainless steel is used to avoid plasticizers.

oratory-reared and laboratory-bred amphibians as research animals for the scientific community. As of October 1977, the colony is composed of over 10,000 living, postmetamorphic animals. This number is not sufficient to permit distribution of large quantities of adult animals to many investigators. However some adults can be provided on a selected basis, and tadpoles of almost any stage are available throughout the year. Arrangements can also be made for other services as described in a price list which may be obtained on request.

The Amphibian Facility can provide *Xenopus laevis*, *Bombina orientalis*, and *Hyperolius* sp., but its major focus continues to be Northern *Rana pipiens*. It is to be noted that a major realization in the past several years has been that although the life cycles of ranids from more temperate areas seem quite amenable to control as revealed by their culture at the Laboratory for Amphibian Biology, Hiroshima, by Bagnara (personal communication) at Tempe, Arizona, and at the Amphibian Facility, the Northern *R. pipiens* and *R. catesbeiana* are particularly refractory. Progress is being made on both species, however, and we have succeeded in bringing Northern *R. pipiens* to maturity in 8 months, compared with 24 to 36 months in nature, and progress has been reported on work with *R. catesbeiana* (Culley, 1977).

Two commercial concerns, one over three years old (Xenopus Ltd., #60) and one currently becoming established (Ozark Amphibian Center, #39), have adopted Amphibian Facility techniques in their attempts to provide service to the scientific and teaching communities. Thus, although the recommendations of the Sub-Committee on Amphibian Standards, Committee on Standards of the National Research Council (Nace et al., 1974) and the Committee on Herpetological Resources (Wake, et al., 1975) regarding frog culture will not be fully satisfied immediately, there is reason to believe that within the next few years amphibian culture will contribute to the alleviation of the demand for amphibians in a much more significant way than seemed possible six years ago.

Regarding the list of suppliers, it will be noted that no prices or statements concerning the availability or conditions under which these animals can be supplied are included in this list. Also, the species are listed as given by the supplier. We have not attempted to review the taxonomic validity of the terms. The supply and developmental stage of animals from all sources are subject to seasonal and other factors which are generally beyond the control of the supplier. Particularly difficult in recent years has been the problem of transportation. Deterioration in the services of the U.S. Postal Service and vagaries in the use of commercial transport concerns have added to the cost and difficulty of transporting animals. Users are urged to communicate directly with the listed sources to obtain the necessary purchasing or exchange information as well as the details concerning the availability of materials.

Bibliography

Bagnara, J. T., and J. S. Frost. 1977. Leopard frog supply. *Science* 197:106-107.

Cooke, A. S. 1972. Indications of recent changes in status in the British Isles of the frog

(*Rana temporaria*) and the toad (*Bufo bufo*). *J. Zool. Proc. Zool. Soc. London* 167: 161-178.

Culley, D. D. 1977. Culture and management of the laboratory frog. *Lab. Anim.* 5:30-36.

Eldridge, A. L. 1976. *Xenopus laevis.* Laboratory studies of the African clawed frog. NASCO, Fort Atkinson, Wisconsin.

Emmons, M. B. (personal communication).

Fite, K. V. 1973. *The amphibian visual system.* New York: Academic Press.

Frost, J. S., and J. T. Bagnara. 1976. A new species of leopard frog (*Rana pipiens* Complex) from Northwestern Mexico. *Copeia* 1976:332-338.

Gibbs, E. L., G. W. Nace, and M. B. Emmons, 1971. The live frog is almost dead. *BioScience* 21:1027-1034.

Hine, R. L., B. L. Les, B. F. Hellmich, and R. C. Vogt. 1975. Preliminary report of leopard frog (*Rana pipiens*) populations in Wisconsin. Department of Natural Resources Research Report 81. Madison, Wisconsin.

Llinás, R., and W. Precht. 1976. *Frog neurobiology.* New York: Springer-Verlag.

Lofts, B. 1976. *Physiology of the amphibia.* Vol. III. New York: Academic Press.

Malacinski, G. H., and A. J. Brothers. 1974. Mutant genes in the Mexican axolotl. *Science* 184:1142-1147.

McKinnell, R. G. (personal communication).

McKinnell, R. G. 1978. Cloning: Nuclear transplantation in amphibia. Univ. of Minnesota Press.

Müller, Hk. 1976. The frog as an experimental animal. In R. Llinás and W. Precht (eds.), *Frog neurobiology.* New York: Springer-Verlag.

Nace, G. W. 1968. The Amphibian Facility of the University of Michigan. *BioScience* 18: 767-775.

Nace, G. W. 1970. The use of amphibians in biomedical research. In: *Animal models for biomedical research III.* Proceedings of a Symposium. National Academy of Sciences, Washington, D.C. Pp. 103-124.

Nace, G. W. 1976a. Chapter 8: Standards for laboratory amphibians. In: K. V. Fite (ed.), *The amphibian visual system: A multidisciplinary approach.* New York: Academic Press. Pp. 317-327.

Nace. G. W. 1976b. Letter published in catalog of Ann Arbor Biological Supply, Inc. P. 100.

Nace, G. W. 1976c. Caretaker's corner: The frog and other amphibians. *Lab Animal* 5: 52-54.

Nace, G. W. 1977. Breeding amphibians in captivity. *International Zoo Yearbook* 17: 44-50.

Nace, G. W., D. D. Culley, M. B. Emmons, E. L. Gibbs, V. H. Hutchison, and R. G. McKinnell. 1974. Amphibians: Guidelines for the breeding, care and management of laboratory animals. I.L.A.R. (NAS/NRC). Washington, D.C. 150 pp. (Available as ISBN 0-309-00210-X from the Printing and Publishing Office, National Academy of Sciences, 2101 Constitution Avenue, N.W., Washington, D.C. 29418.)

Nace, G. W., C. M. Richards, and G. M. Hazen. 1973. Information control in the Amphibian Facility: The use of *R. pipiens* disruptive patterning for individual identification and genetic studies. *Amer. Zool.* 13:115-137.

Nace, G. W., J. K. Waage, and C. M. Richards. 1971. Sources of amphibians for research. *BioScience* 21:768-773.

Pace, A. E. 1974. Systematic and biological studies of the leopard frogs (*Rana pipiens* complex) of the United States. *Misc. Publ., Mus. Zool. Univ. Mich.* 148:1-40.

Symposium. 1973. The laboratory frog: Acquisition, nurture, and health. *Amer. Zool.* 13:71-143.

Wake, D. B., R. C. Zweifel, H. C. Dessauer, G. W. Nace, E. R. Pianka, G. B. Rabb, R. Ruibal, J. W. Wright, and G. R. Zug. 1975. Report of the Committee on Resources in Herpetology. *Copeia* 1975:391-404.

LIST OF SUPPLIERS

Prepared at the Amphibian Facility, October 1977

Address: Division of Biological Sciences, The University of Michigan, Ann Arbor, Michigan 48109

Telephone: Office 313-764-1471; Facility 313-764-7104

† Some animals from the wild near their place of business

* Some animals obtained from stocks reared or bred on their premises

§ Define their animals by the standard terminology found in Nace et al., 1974.

* §   1. Amphibian Facility
      University of Michigan
      405 S. Fourth Street
      Ann Arbor, Michigan 48109
        Att: Dr. George W. Nace, Director
        Tel: 313-764-7104 or 764-1471

*   2. Ann Arbor Biological Center, Inc.
      6780 Jackson Road
      Ann Arbor, Michigan 48103
        Tel: 313-761-8600

*   3. Axolotl Colony
      Department of Biology
      Indiana University
      Bloomington, Indiana 47401
        Att: Dr. George Malacinski
        Tel: 812-337-1131 or 337-2630

—. Bankston, J. (see #25)

*   4. Dr. J. C. Beetschen
Laboratoire de Biologie General
118 Route de Narbonne
Université Paul Sabatier
Toulouse, Cedex 31077
France
   Tel: /61/81-04-85

—. Benton, J. (see #31)

5. Bio-Aquatics International, Inc.
P.O. Box 442
Rochester, Michigan 48063
   Att: Mr. Patrick McKown or Mr. Samuel DeFazio
   Tel: 313-652-6622

6. Boreal Laboratories, Ltd.
1820 Mattawa Avenue
Mississauga, Ontario L4X 1K6
Canada
   Att: Mr. Paul T. Heron, President
   Tel: 416-279-7300
Cable: "BOREALAB"

—. Boyce, B. (see #26)

7. Bravco
P.O. Box 696
Temple, Texas 76501
   Att: Mr. John Bravenec, Jr., Proprietor
   Tel: 817-773-5757

—. Bunts, L. (see #16)

†   8. Carolina Biological Supply Co.
2700 York Road
Burlington, North Carolina 27215
   Tel: 919-584-0381
Cable: "SQUID BURLINGTON NCAR"
(see also Powell Laboratories)

— Cauble, R. (see #19)

9. Charles P. Chase Co., Inc.
7200 N.W. 46th Street
Miami, Florida 33166
   Tel: 305-592-5700
Cable: "BILCHASE"

† 10. F. Cimon Enterprises
P.O. Box 368
Loretteville, Quebec G2B 1G4
Canada
   Tel: 418-842-6933

* 11. Mr. Gary L. Cole
Route 2
P.O. Box 2159
Oroville, California 95965
   Tel: 916-343-0916

12. College Biological Supply
8857 Mt. Israel Road
Escondido, California 92025
   Tel: 714-745-1445
and also

12a. College Biological Supply
21707 Bothell Way
Bothell, Washington 98011
   Tel: 206-486-0731

† 13. Connecticut Valley Biological Supply Co., Inc.
Valley Road
Southhampton, Massachusetts 01073
   Tel: 413-527-4030

† 14. Thomas Cook Ltd.
2 St. George's Street
P.O. Box 12
Cape Town 8000
Republic of South Africa
   Att: Mr. S. K. Down or Mr. B. Rowlands
   Tel: 22-1311
Telex: 57-7013SA
Cable: "THOMACOOK" Cape Town

* §   15. Dr. Dudley D. Culley
Louisiana State University
School of Forestry and Wildlife Management
Baton Rouge, Louisiana 70803
Tel: 504-388-6051

†   16. Dahl Biological Supply Co.
25 Hegenberger Place
Oakland, California 94621
Att: Mr. Larry Bunts, Proprietor
Tel: 415-562-1586

*   17. Dr. Louis Delanney
4712 N.E. 69th Street
Seattle, Washington 98115
Tel: 206-525-2893

†   18. Mr. Angelo DeRosa
Via Mazzini 37
Arzano 80022
Napoli
Italy
Tel: 081-731-3253

19. East Bay Vivarium
1511 MacArthur Boulevard
Oakland, California 94602
Att: Dr. Ronald L. Cauble, Proprietor
Tel: 415-530-7477

−. Emmons, M. (see #37)

†   20. Fisher Scientific Co.
Educational Materials Division
4901 West LeMoyne Avenue
Chicago, Illinois 60651
Tel: 312-378-7770

*   21. Mr. Peter Fraser
22534 Katzman Avenue
Mt. Clemens, Michigan 48043
Tel: 313-956-6050 (day)
313-779-5926 (night)

† 22. Frogs, Inc.
1070 Vickers — KSB&T Bldg.
Wichita, Kansas 67202
Att: Mr. C. L. Leaderbrand, President
Tel: 316-267-1566

—. Ganes, D. (see #41)

23. T. Gerrard and Co.
Gerrard House
Worthing Road, East Preston
Littlehampton, West Sussex BN16 1AS
England
Tel: Rustington 72071
Telex: 87323-FSI /prefix code: Biology

† 24. Graska Biological Supplies, Inc.
P.O. Box 2322
Oshkosh, Wisconsin 54901
Att: Mr. Gerald A. Graska, President
Tel: 414-233-4568

—. Griffin, J. (see #60)

† 25. Gulf-South Biologicals, Inc.
P.O. Box 417
Ponchatoula, Louisiana 70454
Att: Mr. Jim Bankston
Tel: 504-386-8250

† 26. J. M. Hazen and Co.
Alburg, Vermont 05440
Att: Mr. Brian Boyce
Tel: 802-796-3323

* 27. University of Hiroshima
Laboratory of Amphibian Biology
Faculty of Science
Hiroshima
Japan
Att: Dr. M. Nishioka, Director

*   28. Hubrecht Laboratories
International Embryological Institute
Uppsalalaan 8
Universiteitscentrum "De Uithof"
Utrecht
The Netherlands
     Att: Dr. P. D. Nieuwkoop, Director
     Tel: 030-510211

—. Indiana Univ. (see #3)

29. Interfish, Inc.
3000 Middlebelt
Inkster, Michigan 48141
     Att: Mr. Tom Benson
     Tel: 313-326-5555 or
          313-447-9413 (home)
     Cable: "INTERFISH" Detroit

†   30 Jacques Weil Biological Supply Co.
P.O. Box 125
Rayne, Louisiana 70578
     Att: Mr. Mark Sayles
     Tel: 318-334-3689

*   31. J-D Frog Ranch
RFD #2
Canton, Missouri 63435
     Att: Mr. J. L. Benton, Proprietor
     Tel: 314-288-3824

—. Kawamura, T. (see #27)

*   31a. Korea Frog Farms
Bupyong, P.O. Box 9
Inchon 160-70
Korea
     Att: Mr. Lee Jong Jin, Proprietor

32. Kornatexport-Import
    Zagreb
    Yugoslavia
        Att: Mr. Alexander Gulic
        Tel: 041/424-203 or 423-249
        Cable: "KORNATEXPORT ZAGREB"
        Telex: 21340 yu kornat

—. Lantz, L. (see #35)

—. Leaderbrand, C. (see #22)

† 33. Jonathan Leakey, Ltd.
    P.O. Box 1141
    Nakuru
    Kenya
        Att: Mr. J. H. E. Leakey
        Tel: Nairobi Radiocall 2003
        Cable: "LEAKEY" Box 1141 Nakuru

† 34. The Wm. A. Lemberger Associates, Ltd.
    P.O. Box 222
    Germantown, Wisconsin 53022
        Att: Ms. C. F. Lemberger
        Tel: 414-677-4000

—. Malacinski, G. (see #3)

† 35. Midwest Reptile and Animal Sales, Inc.
    P.O. Box 239
    Ft. Wayne, Indiana 46801
        Att: Mr. Larry A. Lantz, President
        Tel: 219-743-1525
        Cable: "MIDREP"

† 36. Mogul-ED
    P.O. Box 2482
    Oshkosh, Wisconsin 54901
        Att: Mr. Tom Daggett
        Tel: 414-231-8410

—. Mortimore, J. (see #59)

—. Nace, G. (see #1)

† * § 37. Nasco
Science Education Division
901 Janesville Avenue
Ft. Atkinson, Wisconsin 53538
   Att: Mr. Marvin B. Emmons, Director
   Tel: 800-558-9595
Telex: 26-5476 NASCO INT FTAN

† * § 38. Nasco Educational Materials Ltd.
58 Dawson Road
Guelph, Ontario N1H 6P9
Canada
   Att: Mr. Alan Baker, Science Manager
   Tel: 519-836-3330
Telex: 069-56588

—. Nieuwkoop, P. (see #28)

—. Nishioka, M. (see #27)

* §   39. Ozark Amphibian Center
P.O. Box 1084
Ozark, Alabama 36360
   Att: Mr. Dan Golda

40. Parco Scientific Co.
316 Youngstown-Kingsville Road, S.E.
Vienna, Ohio 44473
   Tel: 216-856-2368

—. Pawley, R. (see #62)

*   41. The Pond Life Research Co.
P.O. Box 961
Grass Valley, California 95945
   Att: Mr. Dean Ganes, Proprietor
   Tel: 916-272-2224

†   42. The Powell Laboratories
P.O. Box 7
Gladstone, Oregon 97027
   Tel: 503-656-1641
(see also Carolina Biological Supply)

—. Richards, C. (see #1)

†   43. Sargent-Welch Scientific Co.
7300 North Linder Avenue
Skokie, Illinois 60076
   Tel: 312-677-0600
Cable: "WELMANCO"

*   44. Mr. Leonard Scales
820 Fountain Avenue
Ann Arbor, Michigan 48103 (business discontinued)
   Tel: 313-665-2968

*   45. Slabaugh Bullfrog Farm
Rt. 3
P.O. Box 59
Popular Bluff, Missouri 63901
   Att: Mr. Leonard Slabaugh
   Tel: 314-785-7517

†   46. Snake Farm
P.O. Box 96
LaPlace, Louisiana 70068
   Att: Mr. Dan Vicknair, Proprietor
   Tel: 504-652-9229

†   47. South African Snake Farm
P.O. Box 6
Fish Hoek 7975
Republic of South Africa
   Att: Mr. John Wood, Proprietor
   Tel: 82-2912
Cable: "VENOMOUS" Fish Hoek, S.A.

  48. Southern Biological Supply Co.
McKenzie, Tennessee 38201
   Att: Mr. A. C. King
   Tel: 901-352-3000

† 49. Southwestern Scientific Supply Co.
   4345 East Irvington Road
   Tucson, Arizona 85714
      Att: Dr. Lee Stackhouse, Proprietor
      Tel: 602-790-7737

† * 50. Stacel
   79290 Argenton-L'Eglise
   Station D'Acclimation et D'Elevage
   Bouille-Saint-Paul Deux-Sevres
   France
      Att: Mr. R. Guibert
      Cable: "STACEL ARGENTON EGLISE FRANCE 79"

   Stackhouse, L. (see #49)

† 51. Mr. Charles D. Sullivan
   4700 Humber Drive. A-10
   Nashville, Tennessee 37211
      Tel: 615-832-0958

52. Tropical Animal Distributors, Inc.
   156 N.W. 37th Street
   Miami, Florida 33127
      Att: Mr. Robinson Rogers, Jr., Proprietor
      Tel: 305-576-0335

53. Turtox Products
   Macmillan Science Co.
   8200 South Hoyne Avenue
   Chicago, Illinois 60620
      Tel: 312-488-4100

† 54. Mr. Paul van den Elzen
   P.O. Box 499
   Stellenbosch
   Republic of South Africa
      Tel: Stellenbosch 2986

   —. Vicknair, D. (see #46)

55. Wards Natural Science Establishment, Inc.
    P.O. Box 1712
    Rochester, New York 14603
      Tel: 716-467-8400
    Cable: "COSMOS" Rochester
    (see also Wards of California)

56. Wards of California
    P.O. Box 1749
    Monterey, California 93940
      Tel: 408-375-7294
    Cable: "COSMOS" Monterey
    (see also Wards Nat. Sci. Estab.)

*   57. Mr. Travis L. Watson
    Rt. 5
    P.O. Box 248W
    Ft. Worth, Texas 76126
      Tel: 817-443-0239 or 921-2333

58. West Jersey Biological Farm
    South Marion Avenue
    Wenonah, New Jersey 08090
      Att: Mr. E. D. Parker, Proprietor
      Tel: 609-468-1776

*   59. Western Scientific Supply Co.
    1705 South River Road
    P.O. Box 681
    West Sacramento, California 95691
      Att: Mr. Joel Mortimore, President
      Tel: 916-371-2705

—. Wood, J. (see #47)

* §   60. Xenopus Ltd.
    151 Frenches Road
    Red Hill, Surrey RH1 2HZ
    England
      Att: Mr. John Griffin, Proprietor
      Tel: Nutfield Ridge /073 782/2687 or
              Redhill /0737/67224
    Telex: 945935/Exporc/

*   61. Zetts Fish Farm and Hatcheries
     Drifting, Pennsylvania 16834
         Att: Mr. Andrew J. Zetts
         Tel: 814-345-5357 or
               814-345-6502 (home)

     62. Zoological Center International
     Hinsdale, Illinois 61257
         Att: Mr. Ray Pawley
         Tel: 312-325-3256

†   63. Zoologische Grosshandlung
     Gutenbergstrasse 30
     P.O. Bose 98
     Holzminden 345
     West Germany
         Att: Mr. Alfred Koch
         Tel: 05531/7150

Animals Available from Specified Suppliers

Anurans (frogs and toads)

BUFONIDAE

Bufo sp. 35, 43, 63

B. alvarius (Colorado river toad) 35, 51

B. americanus (American toad) 5, 13, 20, 30, 37, 51, 55, 56, 59, 60, 62

B. angusticeps 54

B. asper (hill toad) 35

B. blombergi (Blomberg's toad) 5, 35

B. boreas (Western toad) 5, 19, 51, 59, 60

Bufo bufo (common toad) 5, 23, 35, 50, 59, 60

B. calamita (Natterjack) 51

B. cognatus (Great Plains toad) 5, 51

B. compactilus (Texas toad) 51

B. debilis (green toad) 5, 51

B. gariepensis 54

B. guercicus 5

B. marinus (marine or giant toad) 5, 9, 34, 35, 36, 37, 40, 41, 49, 50, 51, 52, 55, 56, 58, 60, 62

B. mauretanicus (Berber toad) 35

B. melanostictus 35

B. pardalis 47, 54

B. poweri (garden toad) 5

B. punctatus (red-spotted toad) 5, 51

B. quercicus (oak toad) 51

B. rangeri 54

B. regularis (African squareback
toad) 33, 35, 47
B. rosei 54
B. terrestris (Southern toad) 5, 35,
51, 60

B. valliceps (Gulf Coast toad) 5, 51
B. viridis (green toad) 5, 35, 60
B. woodhousei (Woodhouse or
Fowler's toad) 5, 13, 30, 35, 51,
60

DENDROBATIDAE

Dendrobates auratus (arrowhead
poison frog) 19

DISCOGLOSSIDAE

Alytes obstetricans (midwife toad)
51, 60
Bombina sp. 63
B. bombina (fire-bellied toad) 18,
19, 35

B. orientalis (bell frog) 1, 27, 60
B. variegata (yellow-bellied toad)
35, 60
Discoglossus pictus (Spanish painted
frog) 35

HYLIDAE

Acris blanchardi (Blanchard's cricket
frog) 51
A. crepitans (Northern cricket frog)
5, 51
A. gryllus (Southern cricket frog) 5,
51
Agalychnis dachnicolor 19, 49
Hyla sp. 16, 36, 43, 49
H. arborea (European treefrog) 18,
35, 50, 51, 60
H. arenicolor (canyon treefrog) 35,
51
H. avivoca (bird-voiced treefrog) 51
H. cadaverina (California treefrog) 51
H. chrysoscylis (Southern grey tree-
frog) 51
H. cinerea (green treefrog) 5, 19, 23,
30, 35, 46, 49, 51, 60
H. crucifer (spring peeper) 5, 13, 51,
60
H. femoralis (pine wood treefrog) 51

H. gratiosa (barking treefrog) 5, 19,
35, 51
H. regilla (Pacific treefrog) 19, 35,
51, 59
H. septentrionalis (Cuban treefrog) 9,
51
H. squirella (squirrel treefrog) 5, 51
H. vasta (giant treefrog) 9
H. versicolor (common or gray tree-
frog) 5, 13, 19, 35, 51, 60
Limnaoedus ocularis (little grass
frog) 51
Phrynohyas venulosa 19
Pseudacris brachyphona (mountain
chorus frog) 51, 60
P. clarki (spotted chorus frog) 5, 51
P. nigrita (chorus frog) 5, 51
P. ornata (ornate chorus frog) 5, 51
P. streckeri (Strecker's chorus frog)
51, 60
P. triseriata 5, 51

LEPTODACTYLIDAE

Ceratophrys calcarata (horned frog) 9, 35

Eleutherodactylus ricordi (green-house frog) 51

Heleophryne sp. 54

MICROHYLIDAE

Breviceps sp. 54

Gastrophrynae carolinensis (Eastern narrowmouth toad) 51

Gastrophrynae olivacea (Western narrowmouth toad) 51

G. carolinensis 5

G. olivacea 5

G. variolosus 5

Kaloula pulchra 19, 35

PELOBATIDAE

Pelobates fuscus 63

Scaphiopus sp. 35

S. bombifrons (plains spadefoot toad) 51, 60

S. couchi (Couch's spadefoot toad) 35 51

S. hammondi (Western spadefoot toad) 51

S. holbrooki (Eastern spadefoot toad) 51, 60

S. hurteri (Hurter's spadefoot toad) 51

PHRYNOMERIDAE

Phrynomerus sp. 35

P. bifasciatus (red-backed frog) 54

PIPIDAE

Pipa pipa (Surinam toad) 35

Xenopus borealis (Kenyan clawed toad) 60

X. gilli 5, 54

X. laevis (South African clawed toad) 1, 2, 5, 14, 16, 21, 23, 28, 29, 33, 36, 37, 38, 39, 40, 44, 47, 49, 50, 51, 54, 60

X. laevis (periodic albinism) (South African clawed toad) 1, 28, 44

X. mulleri 44

X. tropicalis (Nigerian clawed toad) 60

RANIDAE

Arthroleptella sp. 54

Cacosternum capense 54

Ptychadena anchietae (Savanna sharp-nosed frog) 33

P. mascareniensis (green-striped frog) 33

Pyxicephalus adspersus 19, 54

P. delalandii 54

Rana sp. 16, 54, 62

R. adspersa (African bullfrog) 5, 33

R. areolata (crawfish frog) 5, 51

R. *berlandieri* (Rio Grande leopard frog) 1, 51
R. *blairi* (plains leopard frog) 1, 5, 51
R. *capito* (gopher frog) 51
R. *catesbeiana* (bullfrog) 5, 7, 10, 11, 12, 12a, 13, 15, 16, 20, 22, 25, 30, 31, 31a, 34, 35, 36, 37, 38, 40, 41, 43, 45 46, 48 49, 51, 53, 57, 58, 59, 60, 61, 62
R. *catesbeiana* x R. *grylio* 7
R. *clamitans* (green or bronze frog) 5, 24, 34, 35, 36, 37, 38, 46, 51
R. *esculenta* (edible frog) 18, 23, 32, 50, 60, 63
R. *graeca* 18
R. *grylio* (pig frog) 5, 7, 25, 49, 51
R. *heckscheri* (river frog) 5, 51
R. *japonica* 27

R. *nigromaculata* 27
R. *occipitalis* (groove-crowned bull-frog) 33
R. *palustris* (pickerel frog) 5, 35, 51, 60
R. *pipiens* (leopard or grass frog) 1, 5, 6, 8, 10, 12, 12a, 13, 16, 20, 22, 23, 24, 25, 26, 29, 30, 34, 35, 36, 37, 38, 39, 40, 42, 43, 48, 49, 53, 55, 56, 58, 59, 60, 62
R. *ridibunda* (marsh frog) 60
R. *septentrionalis* 5
R. *sylvatica* (wood frog) 5, 13, 35, 51
R. *temporaria* 23, 24, 59, 60, 63
R. *tigrina* 16
R. *utricularia* (sphenocephala) (Southern leopard frog) 1, 5, 51

RHACOPHORIDAE

Chiromantis petersi 35
Hyperolius sp. 1, 33, 35, 54

Kassina senegalensis (African walking frog) 35

Urodeles (salamanders and newts)

AMBYSTOMATIDAE

Ambystoma sp. 5, 19, 34, 43, 48
A. *annulatum* (ringed salamander) 51
A. *cingulatum* (frosted flatwoods sala-mander) 51
A. *jeffersonianum* 5, 35, 51, 60
A. *maculatum* (spotted salamander) 5, 13, 35, 36, 51, 60
A. *mexicanum* (Mexican axolotl) 1, 3, 4, 17, 23, 28, 60

A. *opacum* (marbled salamander) 5, 35, 51, 60
A. *talpoideum* (mole salamander) 51, 60
A. *texanum* (small-mouthed sala-mander) 5, 35, 51
A. *tigrinum* (tiger salamander) 5, 20, 24, 30, 35, 36, 38, 50, 51, 59, 60, 62

AMPHIUMIDAE

Amphiuma sp. (Congo eel) 5, 8, 25, 30, 36, 46, 48, 49, 59
A. *means* (two-toed amphiuma) 51, 60

A. *tridactylum* (three-toed amphiuma) 51

SALAMANDRIDAE

Cynops pyrrhogaster (= Triturus
  pyrrhogaster) (Japanese fire-
  bellied newt) 27
Diemictylus perstriatus (striped
  newt) 51
Notophthalmus viridescens
  (= Triturus or Diemictylus
  viridescens) (red-spotted newt)
  5, 8, 13, 20, 36, 51, 55, 56,
  59, 60, 61
Pleurodeles waltlii (Spanish ribbed
  newt) 4, 28, 50, 51, 60
Salamandra maculosa 63
S. salamandra (fire salamander) 19,
  23, 35, 51, 60, 62
Taricha sp. (California or Western
  newt) 8, 16, 42, 59

T. granulosa (rough-skinned newt)
  51
T. rivularis (red-bellied newt) 51
T. torosa (Western newt) 51
Triturus sp. 48
T. alpestris (alpine newt) 23, 35,
  51, 60, 63
T. cristatus (crested newt) 18, 23,
  50, 51, 60, 62, 63
T. helveticus (palmate newt) 23, 35,
  50, 51
T. italicus 18
T. marmoratus (marbled newt) 35,
  50, 51, 60
T. palmatus 63
T. vulgaris (common or smooth
  newt) 18, 23, 51, 60, 63

SIRENIDAE

Pseudobranchus striatuss (mud or
  dwarf siren) 5, 51
Siren sp. 48, 59

S. intermedia (lesser siren) 51
S. lacertina (greater siren) 5, 51,
  60

Location of Suppliers

A. Northeastern U.S.A. (Conn.,
   Mass., Me., N.H., N.J., N.Y., Pa.,
   R.I., Vt.; Quebec & eastern Cana-
   da) 10, 13, 26, 55, 58, 61
B. Southeastern Coastal States (Del.,
   Fla., Ga., Md., Eastern N.C., S.C.,
   Va.) 8, 9, 52
C. Great Lakes States (Ill., Ind.,
   Iowa, Mich., Minn., Ohio, Wis.;
   Ontario) 1, 2, 3, 5, 6, 20, 21, 24,
   29, 34, 35, 36, 37, 38, 40, 43, 53,
   62
D. Appalachians (Ky., Western N.C.,
   Tenn., W. Va.) 48, 51

E. Lower Mississippi Valley and Gulf
   States (Ala., Ark., La., Miss., Mo.)
   15, 25, 30, 31, 39, 45, 46
F. Northern Plains and Northern
   Rockies (Mont., Neb., N.D., S.D.,
   Wyo.; Alberta, Manitoba, Sask.)
   no listings
G. Southwest and Southern Rockies
   (Ariz., Colo., Kan., N.M., Okla.,
   Tex., Utah) 7, 22, 49, 57
H. Northwestern U.S.A. (Id., Ore.,
   Wash.; British Columbia) 12a, 17,
   42
I. Southern West Coast (Calif., Nev.)
   11, 12, 16, 19, 41, 56, 59

J. Central and South America

no listings

K. Europe and Western Russia 4, 18,

23, 28, 32, 50, 60, 63

L. Africa 14, 33, 47, 54

M. Far East 27, 31a

Stages in the Normal
Development of *Rana pipiens*

The illustrations of stages of *Rana pipiens* development to the onset of feeding are reprinted from W. Shumway. 1940. *Anat. Rec.* 78:139-147, and the descriptions are adapted from this paper.

Descriptions of Stages

1. Unfertilized egg.
2. Gray crescent after fertilization.
3. First cleavage — two blastomeres. ′
4. Second cleavage — four blastomeres.
5. Third cleavage — eight blastomeres.
6. Fourth cleavage — sixteen blastomeres.
7. Fifth cleavage — thirty-two blastomeres.
8. Young blastula.
9. Definitive blastula.
10. Early dorsal lip of blastopore.
11. Mid-gastrula with semicircular blastopore.
12. Definitive gastrula with circular yolk plug.
13. Neural plate stage.
14. Early neurula with neural folds.
15. Neurula that rotates by ciliary action.
16. Neural tube stage, length approximately 2.5-2.7 mm.

17. Tail bud embryo, length approximately 2.8-3.0 mm.
18. Muscular response stage, length approximately 4 mm.
19. Heart beat stage, average length 5 mm.
20. Gill circulation stage. The embryo hatches spontaneously and swims on mechanical stimulation. Average length 6 mm.
21. Cornea transparent with open mouth stage. Average length 7 mm.
22. Tail fin capillary circulation stage, average length 5 mm.
23. Opercular fold stage, average length 9 mm.
24. Closed right operculum stage, average length 10mm.
25. Complete operculum stage, average length 11 mm.

Stage Number	Age—Hours at 18°C		Stage Number	Age—Hours at 18°C		Stage Number	Age—Hours at 18°C
1	0	UNFERTILIZED	7	7.5	32–CELL	13	50	NEURAL PLATE
2	1	GRAY CRESCENT	8	16	MID-CLEAVAGE	14	62	NEURAL FOLDS
3	3.5	TWO-CELL	9	21	LATE CLEAVAGE	15	67	ROTATION
4	4.5	FOUR–CELL	10	26	DORSAL LIP	16	72	NEURAL TUBE
5	5.7	EIGHT–CELL	11	34	MID–GASTRULA	17	84	TAIL BUD
6	6.5	SIXTEEN–CELL	12	42	LATE GASTRULA			

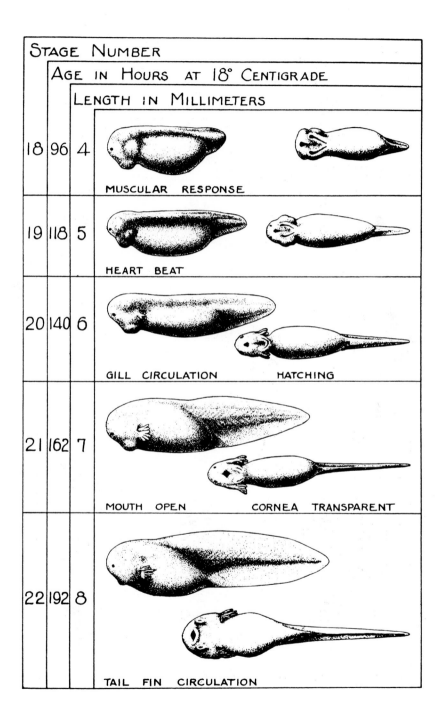

STAGE NUMBER			
	AGE IN HOURS AT 18° CENTIGRADE		
		LENGTH IN MILLIMETERS	
18	96	4	MUSCULAR RESPONSE
19	118	5	HEART BEAT
20	140	6	GILL CIRCULATION HATCHING
21	162	7	MOUTH OPEN CORNEA TRANSPARENT
22	192	8	TAIL FIN CIRCULATION

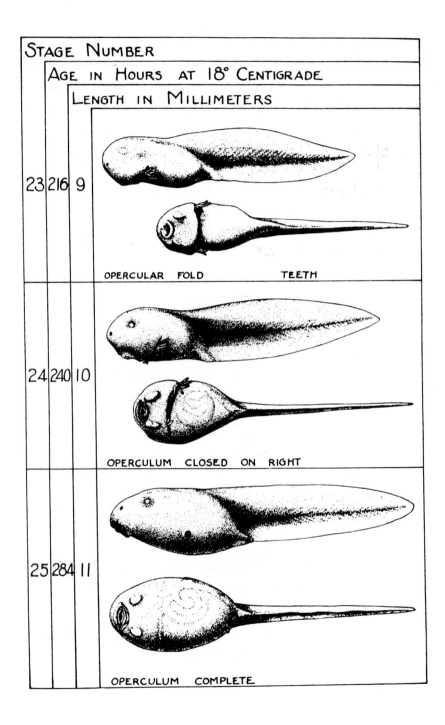

STAGE NUMBER			
	AGE IN HOURS AT 18° CENTIGRADE		
		LENGTH IN MILLIMETERS	
23	216	9	OPERCULAR FOLD TEETH
24	240	10	OPERCULUM CLOSED ON RIGHT
25	284	11	OPERCULUM COMPLETE

BIBLIOGRAPHY

Bibliography

Abrams, G. D. 1969. Diseases in an amphibian colony. In: M. Mizell (ed.), *Biology of amphibian tumors*. New York: Springer-Verlag. Pp. 419–428.

Aimar, C. and M. T. Chalumeau-LeFoulgoc. 1969. Analyse électrophorétique des protéines sériques chez des isojumeaux obtenus par greffe nucléaire dans l'espèce *Pleurodeles waltlii* Michah. (Amphibien Urodèle). *C. R. Hebd. Seances Acad. Sci. Ser. D* 268:368–370.

Aimar, C. and C-L. Gallien. 1972. Etude par la méthode des homogreffes cutanées des réactions immunitaires chez des animaux isogéniques allocytoplasmiques, obtenus par greffes nucléaires intra- et interspécifiques dans la genre *Pleurodeles* (Amphibiens, Urodèles). *C. R. Hebd. Seances Acad. Sci. Ser. D* 274:3019–3022.

Amborski, R. L. and J. C. Glorioso. 1973. The frog revisited. *Science* 181:495.

Anonymous. 1971. Herpes viruses: Elusive carcinogens. *Nature* 231:490–491.

Aronson, L. R. 1943. The sexual behavior of anura. 5. Oviposition in the green frog, *Rana clamitans*, and the bullfrog *Rana catesbeiana*. Am. Mus. Nov. #1224:1–6.

Aydelotte, M. A. 1963. The effects of vitamin A and citral on epithelial differentiation *in vitro*. I. The chick tracheal epithelium. *J. Embryol. Exp. Morphol*. 11:279–291.

Babudieri, B. 1972. Systematics of a leptospira strain isolated from frog. *Experientia* 28:1252–1253.

Bachvorova, R. and E. H. Davidson. 1966. Nuclear activation at the onset of amphibian gastrulation. *J. Exp. Zool*. 163:285–296.

Bagnara, J. T. and J. S. Frost. 1977. Leopard frog supply. *Science* 197:106–107.

Bagnara, J. T. and H. L. Stackhouse. 1973. Observations on Mexican *Rana pipiens*. *Am. Zool*. 13:139–143.

Balinsky, B. I. 1975. *An introduction to embryology*. 4th ed. Philadelphia: W. B. Saunders.

Ballard, W. W. 1976. Problems of gastrulation: Real and verbal. *BioScience* 26:36–39.

Balls, M. and R. H. Clothier. 1974. Spontaneous tumors in amphibia. *Oncology* 29:501–519.

Baltzer, F. 1952. The behavior of nuclei and cytoplasm in amphibian interspecific crosses. *Symp. Soc. Exp. Biol*. 6:230–242.

Barth, L. G. and L. J. Barth. 1959. Differentiation of cells of the *Rana pipiens* gastrula in unconditioned medium. *J. Embryol. Exp. Morphol*. 7:210–222.

Barth, L. J. 1964. *Development: Selected topics*. Reading, Mass.: Addison-Wesley.

Battin, W. T. 1959. The osmotic properties of nuclei isolated from amphibian oocytes. *Exp. Cell Res*. 17:59–75.

287

Bautzmann, H. 1927. Über Induktion sekundärer Embryonalanlagen durch Implantation von Organisatoren in isolierte ventrale Gastrulahälften. *Wilhelm Roux' Arch. Entwicklungsmech. Org.* 110:631–642.

Beetschen, J-C. 1952. Extension et limites du pouvoir régénérateur des membres après la métamorphose chez *Xenopus Laevis* Daudin. *Bull. Biol. Fra. Belg.* 86:88–100.

Benbow, R. M. and C. C. Ford. 1975. Cytoplasmic control of nuclear DNA synthesis during early development of *Xenopus laevis:* A cell free assay. *Proc. Nat. Acad. Sci. USA* 72:2437–2441.

Berns, M. W. 1965. Mortality caused by kidney stones in spinach-fed frogs (*Rana pipiens*). *BioScience* 15:297–298.

Berns, M. W. and K. S. Narayan. 1970. An histochemical and ultrastructural analysis of the dermal chromatophores of the variant Ranid blue frog. *J. Morphol.* 132:169–179.

Bideau, M. 1964. Manifestations cytologiques et comportement des noyaux au cours de la greffe nucléaire chez l'Urodèle *Pleurodeles waltlii* Michah. *C. R. Hebd. Seances Acad. Sci.* 259:213–216.

Blackler, A. W. 1970. The integrity of the reproductive cell line in the amphibia. *Curr. Top. Dev. Biol.* 5:71–87.

Bloch-Shtacher, N., Z. Rabinowitz, and L. Sachs. 1972. Chromosomal mechanism for the induction of reversion in transformed cells. *Int. J. Cancer* 9:632–640.

Bluemink, J. G. 1971. Cytokinesis and cytochalasin-induced furrow regression in the first cleavage zygote of *Xenopus laevis*. *Z. Zellforsch. Mikrosk. Anat.* 121:102–126.

Bobrow, M. and K. Madan. 1973. A comparison of chimpanzee and human chromosomes using the Giemsa-11 and other chromosome banding techniques. *Cytogenet. Cell Genet.* 12:107–116.

Bodenstein, D. 1948. The effects of nitrogen mustard on embryonic amphibian development. II. Effects on eye development. *J. Exp. Zool.* 108:93–125.

Boveri, T. 1902. Über mehrpolige Mitosen als Mittel zur Analyse des Zellkerns. *Verh. Phys.-med. Ges. Wurzburg, NF* 35:67–90. English translation by S. Gluecksohn-Waelsch. In: B. H. Willier and J. M. Oppenheimer (eds.), 1964, *Foundations of experimental embryology*. Englewood Cliffs, N.J.: Prentice-Hall. Pp. 74–97.

Boyer, B. C. 1971. Regulative development in a spiralian embryo as shown by cell deletion experiments on the acoel, *Childia*. *J. Exp. Zool.* 176:97–106.

Brachet, A. 1904. Recherches expérimentales sur l'oeuf de *Rana fusca*. *Arch. Biol.* 21:103–160.

Brachet, J. 1972. The role of the egg cortex in early morphogenesis. In: R. Harris, P. Allin, and D. Viza (eds.), *Cell differentiation*. Copenhagen: Munksgaard. Pp. 3–9.

Brachet, J. 1974. *Introduction to molecular embryology*. New York: Springer-Verlag.

Brachet, J. and E. Hubert. 1972. Studies on nucleocytoplasmic interactions during early amphibian development. I. Localized destruction of the egg cortex. *J. Embryol. Exp. Morphol.* 27:121–145.

Brachet, J., H. Denis, and F. de Vitry. 1964. The effects of actinomycin D and puromycin on morphogenesis in amphibian eggs and *Acetabularia mediterranea*. *Dev. Biol.* 9:398–434.

Brachet, J., A. Pays-de Schutter, and E. Hubert. 1975. Studies on maturation in *Xenopus laevis* oocytes. III. Energy production and requirements for protein synthesis. *Differentiation* 3:3–14.

Bragg, A. N. 1938. The organization of the early embryo of *Bufo cognatus* as revealed especially by the mitotic index. *Z. Zellforsch. Mikrosk. Anat.* 28:154–178.

Brandhorst, B. P. and T. Humphreys. 1971. Synthesis and decay rates of major classes of deoxyribonucleic acid like ribonucleic acid in sea urchin embryos. *Biochemistry* 10:877–881.

Braun, A. C. 1974. *The biology of cancer*. Reading, Mass.: Addison-Wesley.

Breckenridge, W. J. 1944. *Reptiles and amphibians of Minnesota*. Minneapolis: University of Minnesota Press.

Briggs, R. 1947. The experimental production and development of triploid frog embryos. *J. Exp. Zool.* 106:237–266.

Briggs, R. 1973. Developmental genetics of the axolotl. In: F. H. Ruddle (ed.), *Genetic mechanisms of development*. New York: Academic Press. Pp. 169–199.

Briggs, R. and G. Cassens. 1966. Accumulation in the oöcyte nucleus of a gene product essential for embryonic development beyond gastrulation. *Proc. Nat. Acad. Sci. USA* 55:1103–1109.

Briggs, R. and R. Grant. 1943. Growth and regression of frog kidney carcinoma transplanted into the tails of permanent and normal tadpoles. *Cancer Res.* 3:613–620.

Briggs, R. and T. J. King. 1952. Transplantation of living nuclei from blastula cells into enucleated frogs' eggs. *Proc. Nat. Acad. Sci. USA* 38:455–463.

Briggs, R. and T. J. King. 1953. Factors affecting the transplantability of nuclei of frog embryonic cells. *J. Exp. Zool.* 122:485–505.

Briggs, R. and T. J. King. 1955. Specificity of nuclear function in embryonic development. In: E. G. Butler (ed.), *Biological specificity and growth*. Princeton, N.J.: Princeton University Press. Pp. 207–228.

Briggs, R. and T. J. King. 1957. Changes in the nuclei of differentiating endoderm cells as revealed by nuclear transplantation. *J. Morphol.* 110:269–311.

Briggs, R., E. U. Green, and T. J. King. 1951. An investigation of the capacity for cleavage and differentiation in *Rana pipiens* eggs lacking "functional" chromosomes. *J. Exp. Zool.* 116:455–500.

Briggs, R., T. J. King, and M. A. DiBerardino. 1960. Development of nuclear-transplant embryos of known chromosome complement following parabiosis with normal embryos. In: S. Ranzi (ed.), *Symposium on germ cells and development*. Milan: A. Baselli. Pp. 441–477.

Briggs, R., J. Signoret, and R. R. Humphrey. 1964. Transplantation of nuclei of various cell types from neurulae of the Mexican axolotl (*Ambystoma mexicanum*). *Dev. Biol.* 10:233–246.

Bromhall, J. D. 1975. Nuclear transplantation in the rabbit egg. *Nature* 258:719–722.

Brothers, A. J. 1976. Stable nuclear activation dependent on a protein synthesised during oogenesis. *Nature* 260:112–115.

Browder, L. W. 1968a. Pigmentation in *Rana pipiens*. I. Inheritance of the speckle mutation. *J. Hered.* 59:163–166.

Browder, L. W. 1968b. *Rana pipiens* speckle heterozygotes. *J. Hered.* 59:308.

Browder, L. W. 1972. Genetic and embryological studies of albinism in *Rana pipiens. J. Exp. Zool.* 180:149–156.

Browder, L. W. 1975. Frogs of the genus *Rana*. In: R. C. King (ed.), *Handbook of genetics*. Vol. 4. New York: Plenum. Pp. 19–33.

Brown, D. D. 1960. Requirement of Mg^{++} for tadpole growth. *Carnegie Inst. Washington, Yearb.* 60:440.

Brown, D. D. 1967. The genes for ribosomal RNA and their transcription during amphibian development. *Curr. Top. Dev. Biol.* 2:47–73.

Brown, D. D. and J. D. Caston. 1962. Biochemistry of amphibian development. I. Ribosome and protein synthesis in early development of *Rana pipiens. Dev. Biol.* 5:412–434.

Brown, D. D. and E. Littna. 1964. Variations in the synthesis of stable RNAs during oogenesis and development of *Xenopus laevis. J. Mol. Biol.* 8:688–695.

Brown, D. D. and E. Littna. 1966. Synthesis and accumulation of DNA-like RNA during embryogenesis of *Xenopus laevis. J. Mol. Biol.* 20:81–94.

Brown, K. T. and D. G. Flaming. 1974. Beveling of fine micropipette electrodes by a rapid precision method. *Science* 185:693–695.

Brown, L. E. 1973. Speciation in the *Rana pipiens* complex. *Am. Zool.* 13:73–79.

Brun, R. and H. R. Kobel. 1972. Des grenouilles métamorphosées obtenues par transplantation nucléaire à partir du prosencéphale et de l'épiderme larvaire de *Xenopus laevis. Rev. Suisse Zool.* 79:961–965.

Brunst, V. V. 1969. Structures of spontaneous and transplanted tumors in the axolotl (*Siredon mexicanum*). In: M. Mizell (ed.), *Biology of amphibian tumors*. New York: Springer-Verlag. Pp. 215–219.

Burgess, A. M. C. 1967. The developmental potentialities of regeneration blastema cell nuclei as determined by nuclear transplantation. *J. Embryol. Exp. Morphol.* 18:27–41.

Burgess, A. M. C. 1974. Genome control and the genetic potentialities of the nuclei of dedifferentiated regeneration blastema cells. In: G. V. Sherbet (ed.), *Neoplasia and cell differentiation*. Basel: Karger. Pp. 106–152.

Burkitt, D. P. 1972. The trail of a virus—a review. In: P. M. Biggs, G. de-Thé, and L. N. Payne (eds.), *Oncogenesis and herpesviruses*. Lyon: International Agency for Research on Cancer. Pp. 345–348.

Burkitt, D. P., A. M. M. Wilson, and D. B. Jelliffe. 1964. Subcutaneous phycomycosis: A review of 31 cases seen in Uganda. *Br. Med. J.* 1:1669–1672.

Burnstock, N. and J. St. L. Philpot. 1959. A functional test of nuclear suspension media, based on reversible enucleation of amoeba. *Exp. Cell Res.* 16:657–672.

Callan, H. G. 1952. A general account of experimental work on amphibian oocyte nuclei. *Symp. Soc. Exp. Biol.* 6:243–255.

Campbell, A. M. 1969. *Episomes*. New York: Harper and Row.

Carlson, J. T. 1976. Nuclear transfer in *Bombina orientalis*. Ph.D. Thesis, Univ. Minnesota.

Casler, E. 1967. Pattern variation in isogenic frogs. *J. Exp. Zool.* 166:121–135.

Caspersson, T., L. Zech, and C. Johansson. 1970. Differential binding of alkylating fluorochromes in human chromosomes. *Exp. Cell Res.* 60:315–319.

Cavazos, F., J. A. Green, D. G. Hall, and F. V. Lucas. 1967. Ultrastructure of the human endometrial glandular cell during the menstrual cycle. *Am. J. Obstet. Gynecol.* 99:833–854.

Chabry, L. 1887. Contribution à l'embryologie normale et tétratologique des ascidies simples. *J. Anat. Physiol. Norm. Pathol. Homme Anim.* 23:167–319.

Chambers, R. and E. L. Chambers. 1961. *Explorations into the nature of the living cell*. Cambridge, Mass.: Harvard Univ. Press.

Chesterman, F. C. 1967. Viruses. In: R. W. Raven and F. J. C. Roe (eds.), *The prevention of cancer*. New York: Appleton-Century Crofts. Pp. 60–68.

Cohen, N. 1971. Amphibian transplantation reactions: A review. *Am. Zool.* 11:193–205.

Collard, W., H. Thornton, M. Mizell, and M. Green. 1973. Virus-free adenocarcinoma of the frog (summer phase tumor) transcribes Lucké tumor herpesvirus-specific RNA. *Science* 181:448–449.

Comandon, J. and P. de Fonbrune. 1934. Méthode pour l'obtention et l'utilisation des micropipettes. *C. R. Seances Soc. Biol.* 116:1353–1356.

Comandon, J. and P. de Fonbrune. 1939a. Ablation du noyau chez une Amibe. Réactions cinétique à la piqûre de l'Amibe normale ou dénucléée. *C. R. Seances Soc. Biol.* 130:740–744.

Comandon, J. and P. de Fonbrune. 1939b. Greffe nucléaire totale, simple ou multiple, chez une Amibe. *C. R. Seances Soc. Biol.* 130:744–748.

Comandon, J. and P. de Fonbrune. 1942a. Greffes nucléaires croisées entre *Amoeba spheronucleus* et l'une de ses variétés colchiciniques. *C. R. Seances Soc. Biol.* 136:746–747.

Comandon, J. and P. de Fonbrune. 1942b. Modifications héréditaires de volume provoquées par l'échange du noyau entre *Amoeba spheronucleus* et ses variétés colchiciniques. *C. R. Seances Soc. Biol.* 136:747–748.

Comings, D. E. and E. Avelino. 1975. Mechanisms of chromosome banding. VII. Interactions of methylene blue with DNA and chromatin. *Chromosoma* 51:365–379.

Conant, R. 1975. *A field guide to reptiles and amphibians*. 2nd ed. Boston: Houghton Mifflin.

Conklin, E. G. 1905. Mosaic development in ascidian eggs. *J. Exp. Zool.* 2:145–223.

Cope, E. D. 1889. *The batrachia of North America*. United States National Museum Bulletin #34. Reprinted 1963, Ashton, Md.: Eric Lundberg.

Copenhaver, W. M. 1926. Experiments on the development of the heart of *Amblystoma punctatum*. *J. Exp. Zool.* 43:321–371.

Cowdry, E. V. 1955. *Cancer Cells*. Philadelphia: W. B. Saunders.

Culley, Jr. D. D. 1976. Culture and management of the laboratory frog. *Lab Anim.* 5(5):30–36.

Curry, H. A. 1936. Über die Entkernung des Tritoneies durch Absaugen des Eifleckes und die Entwicklung des Tritonmerogons *Triton alpestris* ♀ X *Triton cristatus* ♂ . *Wilhelm Roux' Arch. Entwicklungsmech. Org*. 134:694–715.

Curtis, A. S. G. 1965. Cortical inheritance in the amphibian *Xenopus laevis:* Preliminary results. *Arch. Biol*. 76:523–546.

Curtis, S. K. and E. P. Volpe. 1971. Modification of responsiveness to allografts in larvae of the leopard frog by thymectomy. *Dev. Biol*. 25:177–197.

Dalton, H. C. and Z. P. Krassner. 1959. Role of genetic pituitary differences in larval axolotl pigment development. In: M. Gordon (ed.), *Pigment cell biology*. New York: Academic Press. Pp. 51–61.

Dasgupta, S. 1962. Induction of triploidy by hydrostatic pressure in the leopard frog, *Rana pipiens. J. Exp. Zool*. 151:105–121.

Dasgupta, S. 1969. Developmental capacity of nuclei derived from regenerating blastema of the Mexican axolotl, *Ambystoma mexicanum. Am. Zool*. 9:600.

Dasgupta, S. 1970. Developmental potentialities of blastema cell nuclei of the Mexican axolotl. *J. Exp. Zool*. 175:141–148.

Davidson, E. H. 1976. *Gene activity in early development*. 2nd Ed., New York: Academic Press.

Davidson, E. H., M. Crippa, and A. E. Mirsky. 1968. Evidence for the appearance of novel gene products during amphibian blastulation. *Proc. Nat. Acad. Sci. USA* 60:152–159.

Dawe, C. J., J. Whang-Peng, W. D. Morgan, E. C. Hearnn, and T. Knutsen. 1971. Epithelial origin of polyoma salivary tumors in mice: Evidence based on chromosome-marked cells. *Science* 171:394–397.

Dent, J. N. 1962. Limb regeneration in larvae and metamorphosing individuals of the South African clawed toad. *J. Morphol*. 110:61–77.

de-Thé, G. 1972. Virology and immunology of nasopharyngeal carcinoma: Present situation and outlook—a review. In: P. M. Biggs, G. de-Thé, and L. N. Payne (eds.), *Oncogenesis and herpesviruses*. Lyon: International Agency for Research on Cancer. Pp. 275–284.

Dettlaff, T. A. 1964. Cell divisions, duration of interkinetic states and differentiation in early stages of embryonic development. *Adv. Morphog*. 3:323–362.

Dettlaff, T. A., L. A. Nikitina, and O. G. Stroeva. 1964. The role of the germinal vesicle in oöcyte maturation in anurans as revealed by the removal and transplantation of nuclei. *J. Embryol. Exp. Morphol*. 12:851–873.

Deuchar, E. M. 1972. *Xenopus laevis* and developmental biology. *Biol. Rev. Cambridge Philos. Soc*. 47:37–112.

Deuchar, E. M. 1975. *Xenopus: The South African clawed frog*. London: John Wiley.

DiBerardino, M. A. 1961. Investigations of the germ-plasm in relation to nuclear transplantation. *J. Embryol. Exp. Morphol*. 9:507–513.

DiBerardino, M. A. 1962. The karyotype of *Rana pipiens* and investigation of its stability during embryonic differentiation. *Dev. Biol*. 5:101–126.

DiBerardino, M. A. 1967. Frogs. In: F. H. Wilt and N. K. Wessells (eds.), *Methods in developmental biology*. New York: Thomas Y. Crowell. Pp. 53–74.

DiBerardino, M. A. and N. Hoffner. 1969. Chromosome studies of primary renal carcinoma from Vermont *Rana pipiens*. In: M. Mizell (ed.), *Biology of amphibian tumors*. New York: Springer-Verlag. Pp. 261–278.

DiBerardino, M. A. and N. Hoffner. 1970. Origin of chromosomal abnormalities in nuclear transplants—a reevaluation of nuclear differentiation and nuclear equivalence in amphibians. *Dev. Biol*. 23:185–209.

DiBerardino, M. A. and N. Hoffner. 1971. Development and chromosomal constitution of nuclear-transplants derived from male germ cells. *J. Exp. Zool*. 176:61–72.

DiBerardino, M. A. and N. J. Hoffner. 1975. Nucleo-cytoplasmic exchange of non-histone proteins in amphibian embryos. *Exp. Cell Res*. 94:235–252.

DiBerardino, M. A. and T. J. King. 1965. Transplantation of nuclei from the frog renal adenocarcinoma. II. Chromosomal and histologic analysis of tumor nuclear-transplant embryos. *Dev. Biol.* 11:217–242.

DiBerardino, M. A. and T. J. King. 1967. Development and cellular differentiation of neural nuclear-transplants of known karyotypes. *Dev. Biol.* 15:102–128.

DiBerardino, M. A., T. J. King, and L. Bohl. 1966. Nuclear transplantation of differentiated male germ cells. *Am. Zool.* 6:510.

DiBerardino, M. A., T. J. King, and R. G. McKinnell. 1963. Chromosome studies of a frog renal adenocarcinoma line carried by serial intraocular transplantation. *J. Natl. Cancer Inst.* 31:769–789.

Dickerson, M. C. 1906. *The frog book.* Doubleday, Page. Reprinted 1969, New York: Dover.

Diesch, S. L. and W. F. McCulloch. 1966. Isolation of pathogenic leptospires from waters used for recreation. *Public Health Rep.* 81:299–304.

Driesch, H. 1892a. Entwicklungsmechanische Studien. I. Der Wert der beiden ersten Furchungszellen in der Echinodermenentwicklung. Experimentelle Erzeugung von Teil und Doppelbildungen. *Z. Wiss. Zool.* 53:160–178, 183–184. English translation by L. Mezger, M. and V. Hamburger, and T. S. Hall. In: B. H. Willier and J. M. Oppenheimer (eds.), 1964, *Foundations of experimental embryology.* Englewood Cliffs, N. J.: Prentice-Hall. Pp. 38–50.

Driesch, H. 1892b. Entwicklungsmechanische Studien. IV. Experimentelle Veränderung des Typus der Furchung und ihre Folgen (Wirkungen von Wärmezufuhr und von Druck). *Z. Wiss. Zool.* 55:10–29.

Driesch, H. 1894. *Analytische theorie der organischen entwicklung.* Leipzig: Wilhelm Engelmann.

Driesch, H. 1900. Die isolirten Blastomeren des Echinidenkeimes. *Wilhelm Roux' Arch. Entwicklungsmech. Org.* 10:361–410.

Driesch, H. 1908. *The science and philosophy of the organism.* London: Adam and Charles Black.

Driesch, H. and T. H. Morgan. 1895. Zur Analysis der ersten Entwicklungsstadien des Ctenophoreneies. *Wilhelm Roux' Arch. Entwicklungsmech. Org.* 2:204–224.

DuPasquier, L. 1976. Phylogenesis of the vertebrate immune system. In: F. Melchers and K. Rajewsky (eds.), *The immune system.* Berlin: Springer-Verlag. Pp. 101–115.

DuPasquier, L. and M. R. Wabl. 1977. Transplantation of nuclei from lymphocytes of adult frogs into enucleated eggs: Special focus on technical parameters. *Differentiation,* 8:9–19.

DuPraw, E. J. 1967. The honeybee embryo. In: F. H. Wilt and N. K. Wessells (eds.), *Methods in developmental biology.* New York: Thomas Y. Crowell. Pp. 183–217.

Duryee, W. R. 1938. Isolation of nuclei and non-mitotic chromosome pairs from frog eggs. *Arch. Exp. Zellforsch. Besonders Gewebezuecht.* 19:171–176.

Duryee, W. R. 1950. Chromosomal physiology in relation to nuclear structure. *Ann. N. Y. Acad. Sci.* 50:920–953.

Duryee, W. R. and J. K. Doherty. 1954. Nuclear and cytoplasmic organoids in the living cell. *Ann. N. Y. Acad. Sci.* 58:1210–1230.

El-Badry, H. M. 1963. *Micromanipulators and micromanipulation.* New York: Academic Press.

Elkan, E. 1957. Amphibia IV—*Xenopus laevis* Daudin. In: A. N. Worden and W. Lane-Petter (eds.), *The UFAW handbook on the care and management of laboratory animals.* 2nd ed. Eng.: Kent, Courier Printing and Pub., Pp. 804–815.

Elkan, E. 1960. Some interesting pathological cases in amphibians. *Proc. Zool. Soc., London* 134:275–296.

Ellinger, M. S. 1976. The cell cycle and transplantation of early embryonic nuclei in *Bombina orientalis.* Ph.D. Thesis Univ. Minnesota.

Ellinger, M. S., D. R. King, and R. G. McKinnell. 1975. Androgenetic haploid development produced by ruby laser irradiation of anuran ova. *Radiat. Res.* 62:117–122.

Ellinghausen, Jr. H. C., 1968. Cultural and biochemical characteristics of leptospire from frog kidney. *Bull. Wildlife Dis. Assoc.* 4:41–50.

Elsdale, T. R., M. Fischberg, and S. Smith. 1958. A mutation that reduces nucleolar number in *Xenopus laevis. Exp. Cell Res.* 14:642–643.

Elsdale, T. R., J. B. Gurdon, and M. Fischberg. 1960. A description of the technique for nuclear transplantation in *Xenopus laevis*. *J. Embryol. Exp. Morphol.* 8:437–444.

Emerson, H. and C. Norris. 1905. "Red-leg" — an infectious disease of frogs. *J. Exp. Med.* 7:32–58.

Emery, A. R., A. H. Berst, and K. Kodaira. 1972. Under-ice observations of wintering sites of leopard frogs. *Copeia* 1972:123–126.

Emmons, M. B. 1973. The laboratory frog. *Science* 180:1118.

Endres, H. 1895. Über Anstich- und Schnürversuche an Eiern von Triton taeniatus. *Schles. Ges. Vaterl. Kult., Jahresber.* 73:27–33.

Epstein, M. A. 1972. Virology and immunology of Epstein-Barr virus (EBV) in Burkitt's lymphoma — a review. In: P. M. Biggs, G. de-Thé, and L. N. Payne, (eds.), *Oncogenesis and herpesviruses*. Lyon: International Agency for Research on Cancer. Pp. 261–268.

Estes, R. 1972. Fossil amphibia. *Science* 177:603–604.

Fankhauser, G. 1934. Cytological studies on egg fragments of the salamander *Triton*. IV. The cleavage of egg fragments without the egg nucleus. *J. Exp. Zool.* 67:349–393.

Fankhauser, G. 1945. The effects of changes in chromosome number on amphibian development. *Q. Rev. Biol.* 20:20–78.

Fankhauser, G. 1952. Nucleo-cytoplasmic relations in amphibian development. *Int. Rev. Cytol.* 1:165–193.

Fankhauser, G. 1967. Urodeles. In: F. H. Wilt and N. K. Wessells (eds.), *Methods in developmental biology*. New York: Thomas Y. Crowell. Pp. 85–99.

Fawcett, D. W. 1956. Electron microscope observations on intracellular virus-like particles associated with the cells of the Lucké renal adenocarcinoma. *J. Biophys. Biochem. Cytol.* 2:725–741.

Filoni, S. and V. Margotta. 1971. A study of the regeneration of the cerebellum of *Xenopus laevis* (Daudin) in the larval stages and after metamorphosis. *Arch. Biol.* 82:433–470.

Fischberg, M., J. B. Gurdon, and T. R. Elsdale. 1958a. Nuclear transplantation in *Xenopus laevis*. *Nature* 181:424.

Fischberg, M., J. B. Gurdon, and T. R. Elsdale. 1958b. Nuclear transfer in amphibia and the problem of the potentialities of the nuclei of differentiating tissues. *Exp. Cell Res., Suppl.* 6:161–178.

Flickinger, Jr. R. A. 1949. A study of the metabolism of amphibian neural crest cells during their migration and pigmentation *in vitro*. *J. Exp. Zool.* 112:465–484.

Flickinger, R. A., M. L. Freedman, and P. J. Stambrook. 1967. Generation times and DNA replication patterns of cells of developing frog embryos. *Dev. Biol.* 16:457–473.

Fonbrune, P. de. 1932. Appariel pour fabriquer les instruments de verre destinés aux micromanipulations. *C. R. Hebd. Seances Acad. Sci.* 195:706–707.

Fonbrune, P. de. 1949. *Technique de micromanipulation*. Paris: Masson et Cie.

Ford, C. C. and H. R. Woodland. 1975. DNA synthesis in oocytes and eggs of *Xenopus laevis* injected with DNA. *Dev. Biol.* 43:189–199.

Ford, P. J. 1976. Control of gene expression during differentiation and development. In: C. F. Graham and P. F. Wareing (eds.), *The developmental biology of plants and animals*. Philadelphia: W. B. Saunders. Pp. 302–345.

Fraser, L. R. 1971. Physico-chemical properties of an agent that induces parthenogenesis in *Rana pipiens* eggs. *J. Exp. Zool.* 177:153–172.

Freed, J. J. and S. J. Cole. 1961. Chromosome studies on haploid and diploid cell cultures from *Rana pipiens*. *Am. Soc. Cell Biol.* 1:62.

Freedman, V. H. and S. I. Shin. 1974. Use of the mouse mutant *nude* for studies of cellular tumorigenicity and genetic variations in cultured somatic cells. *Genetics* 77:s23–s24.

Frenkel, N., B. Roizman, E. Cassai, and A. J. Nahmias. 1972. A DNA fragment of Herpes simplex 2 and its transcription in human cervical cancer tissue. *Proc. Nat. Acad. Sci. USA* 69:3784–3789.

Freund, H. 1966. The Greenough stereodouble microscope. An interesting novelty in the micro-surgical field and in intravital microscopy. *Sci. Tech. Info. (Leitz, Wetzlar)* 1:141–142.

Gallien, C. L. and C. Aimar. 1974. Progeny of isogenic males and females obtained by nuclear graft in *Pleurodeles*. *J. Hered.* 65:78–80.

Gallien, L. and M. Durocher. 1957. Table chronologique du développement chez *Pleurodeles waltlii* Michah. *Bull. Biol. Fra. Belg.* 91:97–114.

Gallien, L., B. Picheral, and J-C. Lacroix. 1963a. Transplantation de noyaux triploides dans l'oeuf du Triton *Pleurodeles waltlii* Michah. Developpement de larves viables. *C. R. Hebd. Seances Acad. Sci.* 256:2232–2234.

Gallien, L., B. Picheral, and J-C. Lacroix. 1963b. Modifications de l'assortiment chromosomique chez des larves hypomorphes du Triton *Plurodeles waltlii* Michah., obtenues par transplantation de noyaux. *C. R. Hebd. Seances Acad. Sci.* 257:1721–1723.

Galouzo, I. G. and M. M. Rementsova. 1956. *Bull. Inst. Pasteur, Paris* 55: abst. #57 03379.

Gebhardt, B. M. and E. P. Volpe. 1973. Immunity and tolerance to embryonic tail allografts in the leopard frog. *Transplantation* 15:189–194.

Gibbs, E. L. 1963. An effective treatment for red-leg disease in *Rana pipiens*. *Lab. Anim. Care* 13:781–783.

Gibbs, E. L. 1973. *Rana pipiens*: Health and disease — how little we know. *Am. Zool.* 13:93–96.

Gibbs, E. L., T. J. Gibbs, and T. C. Van Dyck. 1966. *Rana pipiens*: Health and disease. *Lab. Anim. Care* 16:142–160.

Gierthy, J. F. and H. Rothstein. 1971. Cell disorganization and stimulation of mitosis in cultured leopard frog lens. *Exp. Cell Res.* 64:170–178.

Goin, C. J. and O. B. Goin. 1962. *Introduction to herpetology*. San Francisco: W. H. Freeman.

Goldstein, L. 1976. Role for small nuclear RNAs in "programming" chromosomal information? *Nature* 261:519–521.

Goode, R. P. 1967. The regeneration of limbs in adult anurans. *J. Embryol. Exp. Morphol.* 18:259–267.

Goodheart, C. R. 1970. Herpesviruses and cancer. *J. Am. Med. Assoc.* 211:91–96.

Gordon, G. and G. M. Malacinski. 1970. An improved microinjection apparatus for biochemical embryology. *Microchem. J.* 15:685–691.

Graham, C. F. 1970. Parthenogenetic mouse blastocysts. *Nature* 226:165–167.

Graham, C. F., K. Arms, and J. B. Gurdon. 1966. The induction of DNA synthesis by frog egg cytoplasm. *Dev. Biol.* 14:349–381.

Granoff, A. 1973. In vitro growth of Lucké tumor cells. *J. Nat. Cancer Inst.* 51:1087.

Grant, P. 1961. The effect of nitrogen mustard (HN₂) on nucleo-cytoplasmic interaction during amphibian development. In: S. Ranzi (ed.), *Symposium on germ cells and development*. Milan: A. Baselli. Pp. 483–502.

Grant, P. and P. M. Stott. 1962. Effect of nitrogen mustard on nucleocytoplasmic interactions. In: M. Brennan and W. L. Simpson (eds.), *Biological interactions in normal and neoplastic growth*. Boston: Little Brown. Pp. 47–73.

Grant, P. and J. F. Wacaster. 1972. The amphibian grey crescent region — a site of developmental information? *Dev. Biol.* 28:454–471.

Gravell, M. 1971. Viruses and renal carcinoma of *Rana pipiens*. X. Comparison of herpestype viruses associated with Lucké tumor-bearing frogs. *Virology* 43:730–733.

Griffiths, I. 1963. The phylogeny of the salientia. *Biol. Rev. Cambridge Philos. Soc.* 38:241–292.

Gromko, M. H., F. S. Mason, and S. J. Smith-Gill. 1973. Analysis of the crowding effect in *Rana pipiens* tadpoles. *J. Exp. Zool.* 186:63–72.

Gross, P. R. and G. H. Cousineau. 1964. Macromolecule synthesis and the influence of actinomycin on early development. *Exp. Cell Res.* 33:368–395.

Gurdon, J. B. 1959. Tetraploid frogs. *J. Exp. Zool.* 141:519–543.

Gurdon, J. B. 1960a. The effects of ultraviolet irradiation on the uncleaved eggs of *Xenopus laevis*. *Q. J. Microsc. Sci.* 101:299–311.

Gurdon, J. B. 1960b. Factors responsible for the abnormal development of embryos obtained by nuclear transplantation in *Xenopus laevis*. *J. Embryol. Exp. Morphol.* 8:327–340.

Gurdon, J. B. 1961. The transplantation of nuclei between two subspecies of *Xenopus laevis*. *Heredity* 16:305–315.

Gurdon, J. B. 1962a. The developmental capacity of nuclei taken from intestinal epithelium cells of feeding tadpoles. *J. Embryol. Exp. Morphol.* 10:622–640.

Gurdon, J. B. 1962b. Adult frogs derived from the nuclei of single somatic cells. *Dev. Biol.* 4:256–273.

Gurdon, J. B. 1962c. The transplantation of nuclei between two species of *Xenopus*. *Dev. Biol.* 5:68–83.

Gurdon, J. B. 1962d. Multiple genetically identical frogs. *J. Hered.* 53:5–9.

Gurdon, J. B. 1963. Nuclear transplantation in amphibia and the importance of stable nuclear changes in promoting cellular differentiation. *Q. Rev. Biol.* 38:54–78.

Gurdon, J. B. 1967a. On the origin and persistence of a cytoplasmic state inducing nuclear DNA synthesis in frogs' eggs. *Proc. Nat. Acad. Sci. USA* 58:545–552.

Gurdon, J. B. 1967b. African clawed frogs. In: F. H. Wilt and N. K. Wessells (eds.), *Methods in developmental biology*. New York: Thomas Y. Crowell Co. Pp. 75–84.

Gurdon, J. B. 1968a. Changes in somatic cell nuclei inserted into growing and maturing amphibian oocytes. *J. Embryol. Exp. Morphol.* 20:401–414.

Gurdon, J. B. 1968b. Transplanted nuclei and cell differentiation. *Sci. Am.* 219:24–35.

Gurdon, J. B. 1968c. Nucleic acid synthesis in embryos and its bearing on cell differentiation. *Essays Biochem.* 4:25–68.

Gurdon, J. B. 1974a. The assembly of an inexpensive microforge. *Lab. Pract.* February 1974:63–64.

Gurdon, J. B. 1974b. *The control of gene expression in animal development*. London: Oxford University Press.

Gurdon, J. B. 1974c. Molecular biology of a living cell. *Nature* 248:772–776.

Gurdon, J. B. and D. D. Brown. 1965. Cytoplasmic regulation of RNA synthesis and nucleolus formation in developing embryos of *Xenopus laevis*. *J. Mol. Biol.* 12:27–35.

Gurdon, J. B. and C. F. Graham. 1967. Nuclear changes during cell differentiation. *Sci. Prog. (Oxford)* 55:259–277.

Gurdon, J. B. and R. A. Laskey. 1970a. The transplantation of nuclei from single cultured cells into enucleate frogs' eggs. *J. Embryol. Exp. Morphol.* 24:227–248.

Gurdon, J. B. and R. A. Laskey. 1970b. Methods of transplanting nuclei from single cultured cells to unfertilized frogs' eggs. *J. Embryol. Exp. Morphol.* 24:249–255.

Gurdon, J. B. and V. Uehlinger. 1966. "Fertile" intestine nuclei. *Nature* 210:1240–1241.

Gurdon, J. B. and H. R. Woodland. 1969. The influence of the cytoplasm on the nucleus during cell differentiation with special reference to RNA synthesis during amphibian cleavage. *Proc. R. Soc. London, Ser. B.* 173:99–111.

Gurdon, J. B., E. M. DeRobertis, and G. Partington. 1976. Injected nuclei into frog oocytes provide a living cell system for the study of transcriptional control. *Nature* 260:116–120.

Gurdon, J. B., R. A. Laskey, and D. R. Reeves. 1975. The developmental capacity of nuclei transplanted from keratinized cells of adult frogs. *J. Embryol. Exp. Morphol.* 34:93–112.

Gurdon, J. B., J. B. Lingrel, and G. Marbaix. 1973. Message stability in injected frog oocytes: Long life of mammalian α and β globin messages. *J. Mol. Biol.* 80:539–551.

Gurdon, J. B., H. R. Woodland, and J. B. Lingrel. 1974. The translation of mammalian globin mRNA injected into fertilized eggs of *Xenopus laevis*. I. Message stability in development. *Dev. Biol.* 39:125–133.

Guttes, E. and S. Guttes. 1963. Control of DNA replication in an organism with synchronous mitosis. In: S. J. Geerts (ed.), *Genetics Today* New York: Macmillan 1:21.

Guyer, M. F. 1907. The development of unfertilized frog eggs injected with blood. *Science* 25:910–911.

Hadorn, E. 1932. Über Organentwicklung und histologische Differenzierung in tranplantierten merogonischen Bastardgeweben (*Triton palmatus* ♀ X *Triton cristatus* ♂). *Wilhelm Roux' Arch. Entwicklungsmech. Org.* 125:495–565.

296   Bibliography

Häggquist, G. 1965. Presentation speech. In: *Nobel lectures physiology or medicine 1922–1941.* Amsterdam: Elsevier. P. 377.

Hamburger, V. and R. Levi-Montalcini. 1949. Proliferation, differentiation and degeneration in the spinal ganglia of the chick embryo under normal and experimental conditions. *J. Exp. Zool.* 111:457–502.

Hamilton, L. 1957. Androgenetic haploids of a toad, *Xenopus laevis, Nature* 179:159.

Hämmerling, J. 1934. Uber Genomwirkungen und Formbildungfähigkeit bei Acetabularia. *Wilhelm Roux' Arch. Entwicklungsmech. Org.* 132:424–462.

Harris, M. 1964. *Cell culture and somatic variation.* New York: Holt, Rinehart and Winston.

Harrison, F. W., J. Zambernard, and R. R. Cowden. 1975. Fluorescent cytochemistry of calid and algid normal and Lucké tumor-bearing kidneys. *Acta Histochem.* 54:295–306.

Harvey, E. B. 1940. A comparison of the development of nucleate and non-nucleate eggs of *Arbacia punctulata. Biol. Bull. (Woods Hole, Mass.)* 79:166–187.

Hay, E. D. 1959. Electron microscopic observations of muscle dedifferentiation in regenerating *Amblystoma* limbs. *Dev. Biol.* 1:555–585.

Hay, E. D. 1966. *Regeneration.* New York: Holt, Rinehart and Winston.

Hay, E. D. and J. B. Gurdon. 1967. Fine structure of the nucleolus in normal and mutant *Xenopus* embryos. *J. Cell Sci.* 2:151–162.

Hellman, A., M. N. Oxman, and R. Pollack. 1973. *Biohazards in biological research.* New York: Cold Spring Harbor Laboratory.

Hennen, S. 1965. Nucleocytoplasmic hybrids between *Rana pipiens* and *Rana palustris.* I. Analysis of the developmental properties of the nuclei by means of nuclear transplantation. *Dev. Biol.* 11:243–267.

Hennen, S. 1967. Nuclear transplantation studies of nucleocytoplasmic interactions in amphibian hybrids. In: L. Goldstein (ed.), *The control of nuclear activity.* Englewood Cliffs, N.J.: Prentice Hall. Pp. 353–375.

Hennen, S. 1970. Influence of spermine and reduced temperature on the ability of transplanted nuclei to promote normal development in eggs of *Rana pipiens. Proc. Nat. Acad. Sci. USA* 66:630–637.

Hennen, S. 1972. Morphological and cytological features of gene activity in an amphibian hybrid system. *Dev. Biol.* 29:241–249.

Hennen, S. 1973. Competence tests of early amphibian gastrula tissue containing nuclei of one species (*Rana palustris*) and cytoplasm of another (*Rana pipiens*). *J. Embryol. Exp. Morphol.* 29:529–538.

Hennen, S. 1974. Back transfer of late gastrula nuclei of nucleocytoplasmic hybrids. *Dev. Biol.* 36:447–451.

Hertwig, O. 1893. Über den Werth der ersten Furchungszellen für die Organbildung des Embryo. *Arch. Mikroskop. Anat. Entwicklungsmech.* 42:662–807.

Hoffner, N. J. and M. A. DiBerardino. 1977. The acquisition of egg cytoplasmic nonhistone proteins by nuclei during nuclear reprogramming. *Exp. Cell Res.* 108:421–427.

Holliday, R. and J. E. Pugh. 1975. DNA modification mechanisms and gene activity during development. *Science* 187:226–232.

Holtfreter, J. 1931. Über die aufzucht isolierter Teile des Amphibienkeimes. II. Züchtung von keimen und Keimteilen in Salzlösung. *Wilhelm Roux' Arch. Entwicklungsmech. Org.* 124:404–466.

Holtfreter, J. 1954. Observations on the physico-chemical properties of isolated nuclei. *Exp. Cell Res.* 7:95–102.

Hörstadius, S. 1973. *Experimental embryology of echinoderms.* Oxford: Clarendon Press.

Hörstadius, S. and A. Wolsky. 1936. Studien über die Determination der Bilateralsymmetrie des jungen Seeigelkeimes. *Wilhelm Roux' Arch. Entwicklungsmech. Org.* 135:69–113.

Hsu, C. C. and R. G. McKinnell. In prep. The effect of thymectomy on allografts of Lucké renal adenocarcinoma to tadpoles of *Rana pipiens.*

Hsu, T. C., W. Schmid, and E. Stubblefield. 1964. DNA replication sequences in higher animals. In: M. Locke (ed.), *The role of chromosomes in development*. New York: Academic Press. Pp. 83–112.

Hull, T. G. 1955. *Diseases transmitted from animals to man*. 4th ed. Springfield, Ill.: Charles C. Thomas.

Humphreys, T. 1973. RNA and protein synthesis during early animal embryogenesis. In: S. J. Coward (ed.), *Developmental regulation: Aspects of cell differentiation*. New York: Academic Press. Pp. 1–22.

Hunsaker, D. and F. E. Potter. 1960. "Red leg" in a natural population of amphibians. *Herpetologica* 16:285–286.

Hunter, A. S. and F. R. Hunter. 1961. Studies of volume changes in the isolated amphibian germinal vesicle. *Exp. Cell Res.* 22:609–618.

Illmensee, K. 1973. The potentialities of transplanted early gastrula nuclei of *Drosophilia melanogaster*. Production of their imago descendents by germ-line transplantation. *Wilhelm Roux' Arch. Entwicklungsmech. Org.* 171:331–343.

Illmensee, K. and B. Mintz. 1976. Totipotency and normal differentiation of single teratocarcinoma cells cloned by injection into blastocysts. *Proc. Nat. Acad. Sci. USA* 73:549–553.

Illmensee, K., A. P. Mahowald, and M. R. Loomis. 1976. The ontogeny of germ plasm during oogenesis in Drosophilia. *Dev. Biol.* 49:40–65.

Inger, R. F. 1967. The development of a phylogeny of frogs. *Evolution* 21:369–384.

Jacobson, A. G. 1967. Amphibian cell culture, organ culture, and tissue dissociation. In: F. H. Wilt and N. Wessells (eds.), *Methods in developmental biology*. New York: Thomas Y. Crowell. Pp. 531–542.

Jeon, K. W. 1970. Micromanipulation of amoeba nuclei. *Methods Cell Physiol.* 4:179–194.

John, K. R. and D. Fenster. 1975. The effects of partitions on the growth rates of crowded *Rana pipiens* tadpoles. *Am. Midl. Nat.* 93:123–130.

Johnson, K. E. 1974. Gastrulation and cell interactions. In: J. Lash and J. R. Whittaker (eds.), *Concepts of development*. Stamford, Conn.: Sinauer. Pp. 128–148.

Joiner, G. N. and G. D. Abrams. 1967. Experimental tuberculosis in the leopard frog. *J. Am. Vet. Med. Assoc.* 151:942–949.

Jones, K. W. and T. R. Elsdale. 1963. The culture of small aggregates of amphibian embryonic cells *in vitro*. *J. Embryol. Exp. Morphol.* 11:135–154.

Kaplan, H. M. and L. Licht. 1955. Evaluation of chemicals used in control and treatment of diseases of fish and frogs caused by *Pseudomonas hydrophila*. *Am. J. Vet. Res.* 16:342–344.

Kaplan, H. M. and J. G. Overpeck. 1964. Toxicity of halogenated hydrocarbon insecticides for the frog *Rana pipiens*. *Herpetologica* 20:163–169.

Kass, L. R. 1972. New beginnings in life. In: M. P. Hamilton (ed.), *The new genetics and the future of man*. Grand Rapids, Mich.: W. B. Eerdmans. Pp. 15–63.

Kauffeld, C. F. 1937. The status of the leopard frogs, *Rana brachycephala* and *Rana pipiens*. *Herpetologica* 1:84–87.

Kaufman, R. H. and W. E. Rawls. 1974. Herpes genitalis and its relationship to cervical cancer. *Ca.* 24:258–265.

Kawamura, T. 1939. Artificial parthenogenesis in the frog. I. Chromosome numbers and their relation to cleavage histories. *J. Sci. Hiroshima Univ., Ser. B, Div. 1* 6:115–218.

Kawamura, T. 1962. On the names of some Japanese frogs. *J. Sci. Hiroshima Univ., Ser. B, Div. 1* 20:181–193.

Kawamura, T. and M. Nishioka. 1963a. Reciprocal diploid nucleo-cytoplasmic hybrids between two species of Japanese pond frogs and their offspring. *J. Sci. Hiroshima Univ., Ser. B, Div. 1* 21:65–84.

Kawamura, T. and M. Nishioka. 1963b. Reproductive capacity of an amphidiploid male produced by nuclear transplantation in amphibians. *J. Sci. Hiroshima Univ., Ser. B, Div. 1* 21:1–13.

Kawamura, T. and M. Nishioka. 1963c. Nucleocytoplasmic hybrid frogs between two species of Japanese brown frogs and their offspring. *J. Sci. Hiroshima Univ., Ser. B, Div. 1* 21:107–134.

Kawamura, T. and M. Nishioka. 1972. Viability and abnormalities of the offspring of nucleo-cytoplasmic hybrids between *Rana japonica* and *Rana ornativentris*. *Sci. Rep. Lab. Amphibian Biol., Hiroshima Univ.* 1:95–209.

Kedes, L. H. and P. R. Gross. 1969. Identification in cleaving embryos of three RNA species serving as templates for the synthesis of nuclear proteins. *Nature* 223:1335–1339.

King, M. C. 1974. Evolution at two levels: Molecular similarities and biological differences between humans and chimpanzees. *Am. J. Human Genet.* 26:49A.

King, T. J. 1966. Nuclear transplantation in amphibia. *Methods Cell Physiol.* 2:1–36.

King, T. J. 1967. Amphibian nuclear transplantation. In: F. H. Wilt and N. K. Wessells (eds.), *Methods in developmental biology*. New York: Thomas Y. Crowell. Pp. 737–751.

King, T. J. and R. Briggs. 1953. The transplantability of nuclei of arrested hybrid blastulae (*R. Pipiens* ♀ × *R. catesbeiana* ♂). *J. Exp. Zool.* 123:61–78.

King, T. J. and R. Briggs. 1954. Transplantation of living nuclei of late gastrulae into enucleated eggs of *Rana pipiens*. *J. Embryol. Exp. Morphol.* 2:73–80.

King, T. J. and R. Briggs. 1955. Changes in the nuclei of differentiating gastrula cells, as demonstrated by nuclear transplantation. *Proc. Nat. Acad. Sci. USA* 41:321–325.

King, T. J. and R. Briggs. 1956. Serial transplantation of embryonic nuclei. *Cold Spring Harbor Symp. Quant. Biol.* 21:271–290.

King, T. J. and M. A. DiBerardino. 1965. Transplantation of nuclei from the frog renal adenocarcinoma. I. Development of tumor nuclear-transplant embryos. *Ann. N. Y. Acad. Sci.* 126:115–126.

King, T. J. and R. G. McKinnell. 1960. An attempt to determine the developmental potentialities of the cancer cell nucleus by means of transplantation. In: *Cell physiology of neoplasia*. Austin: University of Texas Press. Pp. 591–617.

Kirby, D. R. S. 1971. The transplantation of mouse eggs and trophoblast to extrauterine sites. In: J. C. Daniel, Jr. (ed.), *Methods in mammalian embryology*. San Francisco: W. H. Freeman. Pp. 146–156.

Klein, G. 1972. A summing up. In: P. M. Biggs, G. de-Thé, and L. N. Payne (eds.), *Oncogenesis and herpesviruses*. Lyon: International Agency for Research on Cancer. Pp. 501–515.

Kobel, H. R., R. B. Brun, and M. Fischberg. 1973. Nuclear transplantation with melanophores, ciliated epidermal cells, and the established cell line A-8 in *Xenopus laevis*. *J. Embryol. Exp. Morphol.* 29:539–547.

Kolata, G. B. 1975. Evolution of DNA: Changes in gene regulation. *Science* 189:446–447.

Kopac, M. J. 1964. Micromanipulators: Principles of design, operation, and application. *Phys. Tech. Biol. Res.* 2:191–233.

Kourany, M., C. W. Myers, and C. R. Schneider. 1970. Panamanian amphibians and reptiles as carriers of *Salmonella*. *Am. J. Trop. Med.* 19:632–638.

Kulp, W. L. and D. G. Borden. 1942. Further studies on *Proteus hydrophilus*, the etiological agent in "red leg" disease of frogs. *J. Bacteriol.* 44:673–685.

Lacalli, T. C. and A. B. Acton. 1972. An inexpensive laser microbeam. *Trans. Am. Microsc. Soc.* 91:236–238.

Laskey, R. A. and J. B. Gurdon. 1970. Genetic content of adult somatic cells tested by nuclear transplantation from cultured cells. *Nature* 228:1332–1334.

Legname, C. R. and F. D. Barbieri. 1968. Effect of medium on transplantability of nuclei of *Bufo arenarum* embryonic cells. *Experientia* 24:842.

Lehman, H. E. 1955. On the development of enucleated *Triton* eggs with an injected blastula nucleus. *Biol. Bull. (Woods Hole, Mass.)* 108:138–150.

Lehman, H. E. 1957. Nuclear transplantation, a tool for the study of nuclear differentiation. In: A. Tyler, R. C. von Borstel, and C. B. Metz (eds.). *The beginnings of embryonic development*. Washington, D.C.: Amer. Assn. Advance. Sci. Pp. 201–230.

Levine, N. D. and R. R. Nye. 1976. *Toxoplasma ranae* sp. n. From the leopard frog *Rana pipiens* Linnaeus. *J. Protozool.* 23:488–490.

Levine, N. D. and R. R. Nye. 1977. A survey of blood and other tissue parasites of leopard frogs *Rana pipiens* in the United States. *J. Wildl. Dis.* 13:17–23.

Loeb, J. 1894. Uber eine einfache Methode, zwei order mehr zusammengewachsene Embryonen aus einem Ei hervorzubringen. *Pflüegers Arch.* 55:525–530.

Lopashov, G. V. 1945. Experimental study on potencies of nuclei from newt blastulae by means of transplantation. *Ref. Rab. Biol. Otd. Akad. Nauk. SSSR* 88–89.

Lorch, I. J. and J. F. Danielli. 1950. Transplantation of nuclei from cell to cell. *Nature* 166:329–330.

Lucké, B. 1934. A neoplastic disease of the kidney of the frog, *Rana pipiens. Am. J. Cancer* 20:352–379.

Lucké, B. 1939. Characteristics of frog carcinoma in tissue culture. *J. Exp. Med.* 70:269–276.

Lucké, B. and H. Schlumberger. 1940. The effect of temperature on the growth of frog carcinoma. I. Direct microscopic observations on living intraocular transplants. *J. Exp. Med.* 72:321–330.

Lunger, P. D. 1964. The isolation and morphology of the Lucké frog kidney virus. *Virology* 24:138–145.

Lunger, P. D. 1966. Amphibia-related viruses. *Adv. Virus Res.* 12:1–33.

Lützeler, I. E. and G. M. Malacinski. 1974. Modulations in the electrophoretic spectrum of newly synthesized protein in early Axolotl (*Ambystoma mexicanum*) development. *Differentiation* 2:287–297.

Lynch, J. D. 1973. The transition from archaic to advanced frogs. In: J. L. Vial (ed.), *Evolutionary biology of the anurans.* Columbia: University of Missouri Press. Pp. 133–182.

Mahowald, A. P. and S. Hennen. 1971. Ultrastructure of the "germ plasm" in eggs and embryos of *Rana pipiens. Dev. Biol.* 24:37–53.

Malacinski, G. M. 1971. Genetic control of qualitative changes in protein synthesis during early amphibian (Mexican axolotl) embryogenesis. *Dev. Biol.* 26:442–451.

Malacinski, G. M. and A. J. Brothers. 1974. Mutant genes in the Mexican axolotl. *Science* 184:1142–1147.

Mangold, O. 1923. Transplantationsversuche zur Frage der Spezifität und der Bildung der Keimblätter bei Triton. *Wilhelm Roux' Arch. Entwicklungsmech. Org.* 100:198–301.

Mangold, O. and F. Seidel. 1927. Homoplastische und heteroplastische verschmelzung ganzer Tritonkeime. *Wilhelm Roux' Arch. Entwicklungsmech. Org.* 111:593–665.

Manning, D. D., N. D. Reed, and C. F. Shaffer. 1973. Maintenance of skin xenografts of widely divergent phylogenetic origin on congenitally athymic (nude) mice. *J. Exp. Med.* 138:488–494.

Manson-Bahr, P. H. 1966. *Manson's Tropical Diseases.* Baltimore: Williams and Wilkins.

Markert, C. L. 1968. Neoplasia: A disease of cell differentiation. *Cancer Res.* 28:1908–1914.

Marks, J. and H. Schwabacher. 1965. Infection due to *Mycobacterium xenopei. Br. Med. J.* 1:32–33.

Matsumoto, L. H. and S. Dasgupta. 1974. Cytoplasmic alteration of an established pattern of *Rana pipiens* chromosomal DNA replication. *Dev. Biol.* 37:28–34.

McAvoy, J. W. and K. E. Dixon. 1974. Nuclear transplantation from specialized and unspecialized gut epithelial cells of adult *Xenopus laevis. J. Exp. Zool.* 189:243–248.

McAvoy, J. W., K. E. Dixon, and J. A. Marshall. 1975. Effects of differences in mitotic activity, stage of cell cycle, and degree of specialization of donor cells on nuclear transplantation in *Xenopus laevis. Dev. Biol.* 45:330–339.

McCulloch, W. F., J. L. Braun, and R. G. Robinson. 1962. Leptospiral meningitis: Report of a case and epidemologic follow-up. *J. Iowa Med. Soc.* 52:728–731.

McKinnell, R. G. 1960. Transplantation of *Rana pipiens* (kandiyohi dominant mutant) nuclei to *R. pipiens* cytoplasm. *Am. Nat.* 94:187–188.

McKinnell, R. G. 1962a. Development of *Rana pipiens* eggs transplanted with Lucké tumor cells. *Am. Zool.* 2:430–431.

McKinnell, R. G. 1962b. Intraspecific nuclear transplantation in frogs. *J. Hered.* 53:199–207.

McKinnell, R. G. 1964. Expression of the kandiyohi gene in triploid frogs produced by nuclear transplantation. *Genetics* 49:895–903.

McKinnell, R. G. 1965. Incidence and histology of renal tumors of leopard frogs from the north central states. *Ann. N.Y. Acad. Sci.* 126:85–98.

McKinnell, R. G. 1969. Lucké renal adenocarcinoma: Epidemiological aspects. In: M. Mizell (ed.), *Biology of amphibian tumors.* New York: Springer-Verlag. Pp. 254–260.

McKinnell, R. G. 1972. Nuclear transfer in *Xenopus* and *Rana* compared. In: R. Harris, P. Allin, and D. Viza (eds.), *Cell differentiation.* Copenhagen: Munksgaard. Pp. 61–64.

McKinnell, R. G. 1973a. The Lucké frog kidney tumor and its herpesvirus. *Am. Zool.* 13:97–114.

McKinnell, R. G. 1973b. Nuclear transplantation. In: *Seventh national cancer conference proceedings.* Philadelphia: Lippincott. Pp. 65–72.

McKinnell, R. G. and K. Bachmann. 1965. Quantitative DNA determinations of nuclear transplant triploid *Rana pipiens. Exp. Cell Res.* 39:625–630.

McKinnell, R. G. and D. C. Dapkus. 1973. The distribution of burnsi and kandiyohi frogs in Minnesota and contiguous states. *Am. Zool.* 13:81–84.

McKinnell, R. G. and D. P. Duplantier. 1970. Are there renal adenocarcinoma-free populations of leopard frogs? *Cancer Res.* 30:2730–2735.

McKinnell, R. G. and V. L. Ellis. 1972a. Epidemiology of the frog renal tumour and the significance of tumour nuclear transplantation studies to a viral aetiology of the tumour — a review. In: P. M. Biggs, G. de-Thé, and L. N. Payne (eds.), *Oncogenesis and herpesviruses.* Lyon: International Agency for Research on Cancer. Scientific Publications No. 2. Pp. 183–197.

McKinnell, R. G. and V. L. Ellis. 1972b. Herpesviruses in tumors of postspawning *Rana pipiens. Cancer Res.* 32:1154–1159.

McKinnell, R. G. and B. K. McKinnell. 1967. An extension of the ranges of the burnsi and kandiyohi variants of *Rana pipiens. J. Minn. Acad. Sci.* 34:176.

McKinnell, R. G. and B. K. McKinnell. 1968. Seasonal fluctuation of frog renal adenocarcinoma prevalence in natural populations. *Cancer Res.* 28:440–444.

McKinnell, R. G. and K. S. Tweedell. 1969. Induction of renal tumors in triploid leopard frogs. *Am. Zool.* 9:600.

McKinnell, R. G. and K. S. Tweedell. 1970. Induction of renal tumors in triploid leopard frogs. *J. Nat. Cancer Inst.* 44:1161–1166.

McKinnell, R. G. and J. Zambernard. 1968. Virus particles in renal tumors obtained from spring *Rana pipiens* of known geographic origin. *Cancer Res.* 28:684–688.

McKinnell, R. G., B. A. Deggins, and D. D. Labat. 1969. Transplantation of pluripotential nuclei from triploid frog tumors. *Science* 165:394–396.

McKinnell, R. G., M. F. Mims, and L. A. Reed. 1969. Laser ablation of maternal chromosomes in eggs of *Rana pipiens. Z. Zellforsch. Mikrosk. Anat.* 93:30–35.

McKinnell, R. G., D. J. Picciano, and R. E. Krieg. 1976. Fertilization and development of frog eggs after repeated spermiation induced by human chorionic gonadotropin. *Lab. Anim. Sci.* 26:932–935.

McKinnell, R. G., L. M. Steven, Jr., and E. G. Ellgaard. 1973. Serum electrophoresis of genetic replicate leopard frogs produced by nuclear transplantation. *Differentiation* 1:173–176.

McKinnell, R. G., L. M. Steven, Jr., and D. D. Labat. 1976. Frog renal tumors are composed of stroma, vascular elements, and epithelial cells: What type nucleus programs for tadpoles with the cloning procedure? In: N. Müller-Bérat (ed.), *Progress in differentiation research.* Amsterdam: North Holland. Pp. 319–330.

McKinnell, R. G., V. L. Ellis, D. C. Dapkus, and L. M. Steven, Jr. 1972. Early replication of herpesviruses in naturally occurring frog tumors. *Cancer Res.* 32:1729–1732.

Meins, F., Jr. 1974. Mechanisms underlying tumor transformation and tumor reversal in crown-gall, a neoplastic disease of higher plants. In: T. J. King (ed.), *Developmental aspects of carcinogenesis and immunity.* New York: Academic Press. Pp. 23–39.

Merrell, D. J. 1963. Rearing tadpoles of the leopard frog, *Rana pipiens. Turtox News* 41:263–265.

Merrell, D. J. 1965. The distribution of the dominant burnsi gene in the leopard frog, *Rana pipiens. Evolution* 19:69–85.

Merrell, D. J. 1975. *An introduction to genetics.* New York: W. W. Norton and Co.

Mims, M. F. and R. G. McKinnell. 1971. Laser irradiation of the chick embryo germinal crescent. *J. Embryol. Exp. Morphol.* 26:31–36.

Minowada, J., M. Nonoyama, G. E. Moore, A. M. Rauch, and J. S. Pagano. 1974. The presence of the Epstein-Barr viral genome in human lymphoblastoid B-cell lines and its absence in a myeloma cell line. *Cancer Res.* 34:1898–1903.

Mintz, B. 1964. Synthetic processes and early development in the mammalian egg. *J. Exp. Zool.* 157:85–100.

Miyada, S. 1960. Studies on haploid frogs. *J. Sci. Hiroshima Univ., Ser. B., Div. 1* 19:1–56.

Mizell, M., C. W. Stackpole, and S. Halpern. 1968. Herpes-type virus recovery from "virus-free" frog kidney tumors. *Proc. Soc. Exp. Biol. Med.* 129:808–814.

Moore, A-B. C. 1950. The development of reciprocal androgenetic frog hybrids. *Biol. Bull. (Woods Hole, Mass.)* 99:88–111.

Moore, J. A. 1939. Temperature tolerance and rates of development in the eggs of amphibia. *Ecology* 20:459–478.

Moore, J. A. 1941. Developmental rate of hybrid frogs. *J. Exp. Zool.* 86:405–422.

Moore, J. A. 1942. An embryological and genetical study of *Rana burnsi* Weed. *Genetics* 27:406–416.

Moore, J. A. 1944. Geographic variation in *Rana pipiens* Schreber of eastern North America. *Bull. Am. Mus. Nat. Hist.* 82:349–369.

Moore, J. A. 1947. Studies in the development of frog hybrids. II. Competence of the gastrula ectoderm of *Rana pipiens* ♀ × *Rana sylvatica* ♂ hybrids. *J. Exp. Zool.* 105:349–370.

Moore, J. A. 1955. Abnormal combinations of nuclear and cytoplasmic systems in frogs and toads. *Adv. Genet.* 7:139–182.

Moore, J. A. 1960a. Serial back-transfer of nuclei in experiments involving two species of frogs. *Dev. Biol.* 2:535–550.

Moore, J. A. 1960b. Nuclear transfer of embryonic cells of the amphibia. In: P. M. B. Walker (ed.), *Symposium on new approaches in cell biology.* New York: Academic Press. Pp. 1–14.

Moore, J. A. 1969. Interrelations of the populations of the *Rana pipiens* complex. In: M. Mizell (ed.), *Biology of amphibian tumors.* New York: Springer-Verlag. Pp. 26–34.

Moore, N. W., C. E. Adams, and L. E. A. Rowson. 1968. Developmental potential of single blastomeres of the rabbit egg. *J. Reprod. Fert.* 17:527–531.

Morek, D. M. 1972. An organ culture study of frog renal tumor and its effects on normal frog kidney *in vitro.* Ph.D. Dissertation, University of Notre Dame, Notre Dame, Ind.

Morgan, T. H. 1895. The formation of the fish embryo. *J. Morphol.* 10:419–472.

Morgan, T. H. 1927. *Experimental embryology.* New York: Columbia University Press.

Morrill, J. B., C. A. Blair, and W. J. Larsen. 1973. Regulative development in the pulmonate gastropod *Lymnaea palustris* as determined by blastomere deletion experiments. *J. Exp. Zool.* 183:47–56.

Muggleton, A. and J. F. Danielli. 1958. Ageing in *Amoeba proteus* and *A. discoides* cells. *Nature* 181:1738.

Muggleton, A. and J. F. Danielli. 1968. Inheritance of the "life-spanning" phenomenon in *Amoeba proteus. Exp. Cell Res.* 49:116–120.

Muggleton-Harris, A. L. 1970. Cellular changes occurring with age in the lens cells of the frog (*Rana pipiens*) in reference to the developmental capacity of the transplanted nuclei. *Exp. Gerontol.* 5:227–232.

Muggleton-Harris, A. L. 1971a. Cellular events concerning the developmental potentiality of the transplanted nucleus, with reference to the aging lens cell. *Exp. Gerontol.* 6:279–285.

Muggleton-Harris, A. L. 1971b. Aging factors affecting the ability of adult lens cell nuclei for cleavage and development. *Exp. Gerontol.* 6:461–467.

Muggleton-Harris, A. L. 1972. Aging effects at the cellular level, studied by transferring nuclei from organ cultured lens cells. *Exp. Gerontol.* 7:219–225.

Muggleton-Harris, A. L. and K. Pezzella. 1972. The ability of the lens cell nucleus to promote complete embryonic development through to metamorphosis and its applications to ophthalmic gerontology. *Exp. Gerontol.* 7:427–431.

Nace, G. W. 1968. The amphibian facility of the University of Michigan. *BioScience* 18:767–775.

Nace, G. W., C. M. Richards, and H. Sambuichi. 1966. Establishment of an amphibian facility. *Am. Zool.* 6:547.

Nace, G. W., J. K. Waage, and C. M. Richards. 1971. Sources of amphibians for research. *BioScience* 21:768–773.

Nace, G. W., D. D. Culley, M. B. Emmons, E. L. Gibbs, V. H. Hutchinson, and R. G. McKinnell. 1974. *Amphibians: Guidelines for the breeding, care, and management of laboratory animals.* Washington, D.C.: Institute of Laboratory Animal Resources.

Naegele, R. F., A. Granoff, and R. W. Darlington. 1974. The presence of the Lucké herpesvirus genome in induced tadpole tumors and its oncogenicity: Koch-Henle postulates fulfilled. *Proc. Nat. Acad. Sci. USA* 71:830–834.

Namenwirth, M. 1974. The inheritance of cell differentiation during limb regeneration in the axolotl. *Dev. Biol.* 41:42–56.

Nelsen, O. E. 1953. *Comparative embryology of the vertebrates.* New York: The Blakiston Company.

Nemer, M. and D. T. Lindsay. 1969. Evidence that the s-polysomes of early sea urchin embryos may be responsible for the synthesis of chromosomal histones. *Biochem. Biophys. Res. Commun.* 35:156–160.

Nevo, E. 1968. Pipid frogs from the early cretaceous of Israel and Pipid evolution. *Bull. Mus. Comp. Zool.* 136:255–318.

Nieuwkoop, P. D. and J. Faber. 1967. *Normal table of* Xenopus laevis *(Daudin).* 2nd ed. Amsterdam: North-Holland Pub. Co.

Nikitina, L. A. 1964. Transfers of nuclei from the ectoderm and nervous rudiment of developing embryos of *Bufo bufo, Rana arvalis,* and *Rana temporaria* into enucleated eggs of the same species. *Dokl. Acad. Nauk. SSSR* 156:1468–1471.

Nikitina, L. A. 1969. A study of age changes in nuclei of ectoderm and its derivatives by the method of nuclear transplantation (experiments on anuran embryos). *Tsitologiya* 11:542–553.

Nikitina, L. A. 1972. Transplantation of material from the germinal vesicle to enucleated sturgeon oocytes. *Dokl. Acad. Nauk SSSR* 205:1487–1489.

Nikitina, L. A. 1974. Transplantation of nuclei of growing oocytes into enucleated sturgeon oocytes of definitive size. *Ontogenez* 5:289–293.

Nishioka, M. 1972a. Nucleo-cytoplasmic hybrids between *Rana japonica* and *Rana temporaria temporaria. Sci. Rep. Lab. Amphibian Biol., Hiroshima Univ.* 1:211–243.

Nishioka, M. 1972b. Reciprocal nucleo-cytoplasmic hybrids between *Rana esculenta* and *Rana brevipoda. Sci. Rep. Lab. Amphibian Biol., Hiroshima Univ.* 1:245–257.

Nishioka, M. 1972c. Nucleo-cytoplasmic hybrids between *Rana brevipoda* and *Rana plancyi chosenica. Sci. Rep. Lab. Amphibian Biol., Hiroshima Univ.* 1:259–275.

Olsen, M. W. 1969. Potential uses of parthenogenetic development in turkeys. *J. Hered.* 60:346–348.

Oppenheimer, J. M. 1965. Questions posed by classical descriptive and experimental embryology. In: J. A. Moore (ed.), *Ideas in modern biology.* Garden City, N.Y.: Natural History Press. Pp. 205–227.

Ortolani, G., M. Fischberg, and S. Slatkine. 1966. Nuclear transplantation between two subspecies of *Xenopus laevis (Xenopus laevis laevis* and *Xenopus laevis petersi*). *Acta. Embryol. . Morphol. Exp.* 9:187–202.

Orton, G. L. 1953. The systematics of vertebrate larvae. *Syst. Zool.* 2:63–75.

Pace, A. E. 1974. Systematic and biological studies of the leopard frogs (*Rana pipiens* complex) of the United States. *Misc. Publ. Mus. Zool. Univ. Mich.* 148:1–140.

Pantaleon, J. and R. Rosset. 1964. Sur la présence de *Salmonella* dans les grenouilles destinées à la consommation humaine. *Ann. Inst. Pasteur Lille* 15:225–227.

Pantelouris, E. M. 1968. Absence of thymus in a mouse mutant. *Nature* 217:370–371.

Pantelouris, E. M. and J. Jacob. 1958. Nuclear chimeras in the newt. *Experientia* 14:99.

Papaioannou, V. E., M. W. McBurney, R. L. Gardner, and M. J. Evans. 1976. The fate of teratocarcinoma cells injected into early mouse embryos. In: N. Müller-Bérat (ed.), *Progress in differentiation research.* Amsterdam: North Holland Pub. Co. P. 275.

Papermaster, D. S. and E. Gralla. 1973. Frog health. *Science* 180:10.

Parmenter, C. L. 1933. Haploid, diploid, triploid, and tetraploid chromosome numbers and their origin in parthenogenetically developed larvae and frogs of *Rana pipiens* and *Rana palustris. J. Exp. Zool.* 66:409–453.

Pavlovsky, E. N. 1966. *Natural nidality of transmissible diseases.* Urbana: University of Illinois Press.

Penners, A. 1926. Experimentelle untersuchungen zum Determinationsproblem am Keim non *Tubifex rivulorum. Z. Wiss. Zool.* 127:1–140.

Picheral, B. 1962. Capacitiés des noyoux de cellules endodermiques embryonnaires à organiser un germe viable chez l'urodèle, *Pleurodeles waltlii* Miach. *C. R. Hebd. Seances Acad. Sci. Ser. D* 255:2509–2511.

Pierce, G. B. 1972. Differentiation and cancer. In: R. Harris, P. Allin and D. Viza (eds.), *Cell differentiation.* Copenhagen: Munksgaard. Pp. 109–114.

Pierce, G. B. 1974. The benign cells of malignant tumors. In: T. J. King (ed.), *Developmental aspects of carcinogenesis and immunity.* New York: Academic Press. Pp. 3–22.

Pierce, G. B. 1976. Origin of neoplastic stem cells. In: N. Müller-Bérat (ed.), *Progress in differentiation research.* Amsterdam: North Holland Pub. Co. Pp. 269–273.

Pizzarello, D. J. and A. Wolsky. 1960. Sexual dimorphism in histocompatibility reactions of amphibia to skin homografts and a tentative explanation of their mechanisms. *Ann. N.Y. Acad. Sci.* 87:45–54.

Pogany, G. C. 1970. Pigment polymorphism and blood serum variations in *Rana pipiens. Comp. Biochem. Physiol.* 36:99–102.

Porter, K. R. 1939. Androgenetic development of the egg of *Rana pipiens. Biol. Bull. (Woods Hole, Mass.)* 77:233–257.

Rafferty, K. A., Jr. 1962. Age and environmental temperature as factors influencing development of kidney tumors in uninoculated frogs. *J. Nat. Cancer Inst.* 29:253–265.

Rafferty, K. A., Jr. 1964. Kidney tumors of the leopard frog: A review. *Cancer Res.* 24:169–185.

Rafferty, K. A., Jr. 1965. The cultivation of inclusion associated viruses from Lucké tumor frogs. *Ann. N.Y. Acad. Sci.* 126:3–21.

Rafferty, K. A., Jr. 1969. Mass culture of amphibian cells: Methods and observations concerning stability of cell type. In: M. Mizell (ed.), *Biology of amphibian tumors.* New York: Springer-Verlag. Pp. 52–81.

Rafferty, N. S. 1963. Studies of an injury-induced growth in the frog lens. *Anat. Rec.* 146:299–311.

Raina, A. and J. Jänne. 1970. Do polyamines play a role in the regulation of nucleic acid metabolism? In: J. J. Blum (ed.), *Biogenic amines as physiological regulators.* Englewood Cliffs, N.J.: Prentice-Hall, Inc. Pp. 275–300.

Raina, A. and J. Jänne. 1975. Physiology of the natural polyamines putrescine, spermidine, and spermine. *Med. Biol.* 53:121–147.

Rauber, A. 1886. Personaltheil und germinaltheil des individuum. *Zool. Anz.* 9:166–171.

Reeves, O. R. and R. A. Laskey. 1975. *In vitro* differentiation of a homogeneous cell population — the epidermis of *Xenopus laevis. J. Embryol. Exp. Morphol.* 34:75–92.

Reichenbach-Klinke, H. H. and E. Elkan. 1965. *The principal diseases of lower vertebrates.* London: Academic Press.

Reverberi, G. 1971. Ascidians. In: G. Reverberi (ed) *Experimental embryology of marine and fresh-water invertebrates.* Amsterdam: North Holland. Pp. 507–550.

Reverberi, G. and G. Ortolani. 1962. Twin larvae from halves of the same egg in ascidians. *Dev. Biol.* 5:84–100.

Reynaud, G. 1973. Contribution à l'etude des relations entre soma et germen chez le poulet au moyen d'une technique de transfert des gonocytes primordiaux. These à l'Universite de Provence.

Richards, C. M. 1958. The inhibition of growth in crowded *Rana pipiens* tadpoles. *Physiol. Zool.* 31:138–151.

Richards, C. M., D. T. Tartof, and G. W. Nace. 1969. A melanoid variant in *Rana pipiens. Copeia* 1969:850–852.

Rivers, T. M. 1936. Viruses and Koch's postulates. *J. Bacteriol.* 33:1–12.

Rose, F. L. 1962. A case of albinism in *Rana pipiens* Schreber. *Herpetologica* 18:72.

Rose, S. M. 1949. Transformed cells. *Sci. Am.* 181:22–24.

Rose, S. M. 1970. *Regeneration: Key to understanding normal and abnormal growth and development.* New York: Appleton-Century-Crofts.

Rose, S. M. and H. M. Wallingford. 1948. Transformation of renal tumors of frogs to normal tissue in regenerating limbs of salamanders. *Science* 107:457.

Rose, W. 1950. *The reptiles and amphibians of southern Africa.* Capetown: Maskew Miller.

Rostand, J. 1943. Essai d'inoculation de noyaux embryonnaires dans l'oeuf vierge de grenouille. Parthenogenese ou fecondation? *Rev. Sci.* 81:454–456.

Roux, K. H. and E. P. Volpe. 1974. Expression of histocompatibility loci in the leopard frog. *J. Hered.* 65:341–344.

Roux, W. 1888. Beiträge zur Entwickelungsmechanik des Embryo. Ueber die Künstliche Hervorbringung halber Embryonen durch Zerstörung einer der beiden ersten Furchungskugeln, sowie über die Nachentwickelung (Postgeneration) der Fehlenden Körperhälfte. *Virchows Arch. Pathol. Anat. Physiol.* 114:113–153; Resultate 289–291. English translation by H. Laufer. In: B. H. Willier and J. M. Oppenheimer (eds.), *Foundations of experimental embryology.* Englewood Cliffs, N.J.: Prentice-Hall, Inc. Pp. 2–37.

Roux. W. 1892. Ziele und Wege der Entwickelungsmechanik. *Ergeb. Anat. Entwicklungsgesch.* 2:415–445.

Ruben, L. N. 1956. The effects of implanting anuran cancer into regenerating adult urodele limbs. I. Simple regenerating systems. *J. Morphol.* 98:389–403.

Rugh, R. 1934. Induced ovulation and artificial fertilization in the frog. *Biol. Bull. (Woods Hole, Mass.)* 66:22–29.

Rugh, R. 1962. *Experimental embryology: Techniques and procedures.* 3rd ed. Minneapolis: Burgess.

Rugh, R. 1968. *The mouse: Its reproduction and development.* Minneapolis: Burgess.

Rush, H. G., M. R. Anver, and E. S. Beneke. 1974. Systemic chromomycosis in *Rana pipiens. Lab. Anim. Sci.* 24:646–655.

Russell, F. H. 1898. An epidemic, septicemic disease among frogs due to the *Bacillus hydrophilus fuscus. J. Am. Med. Assoc.* 30:1442–1449.

Sabin, A. B. and G. Tarro. 1973. Herpes simplex and herpes genitalis viruses in etiology of some human cancers. *Proc. Nat. Acad. Sci. USA* 70:3225–3229.

Sachs, M. I. and E. Anderson. 1970. A cytological study of artificial parthenogenesis in the sea urchin, *Arbacia punctulata. J. Cell Biol.* 47:140–158.

Sambuichi, H. 1957a. The roles of the nucleus and the cytoplasm in development. I. An intersubspecific hybrid frog, developed from a combination of *Rana nigromaculata ni-*

gromaculata cytoplasm and a diploid nucleus of *Rana nigromaculata brevipoda. J. Sci. Hiroshima Univ., Ser. B, Div. 1* 17:33–41.

Sambuichi, H. 1957b. The roles of the nucleus and the cytoplasm in development. II. Transplantation of blastula cell nuclei into the enucleated overripe eggs of the pond frog, *Rana nigromaculata nigromaculata Hallowell J. Sci. Hiroshima Univ., Ser. B, Div. 1* 17:43–46.

Sambuichi, H. 1959. Production of polyploids by means of transplantation of nuclei in frog eggs. *J. Sci. Hiroshima Univ., Ser. B, Div. 1* 18:39–43.

Sambuichi, H. 1961. The roles of the nucleus and the cytoplasm in development. III. Diploid nucleo-cytoplasmic hybrids, derived from *Rana nigromaculata brevipoda* cytoplasm and *Rana nigromaculata nigromaculata* nuclei. *J. Sci. Hiroshima Univ., Ser. B, Div. 1* 20:1–15.

Sambuichi, H. 1964. Effects of radiation on frog eggs. I. The development of eggs transplanted with irradiated nuclei or cytoplasm. *Jpn. J. Genet.* 39:259–267.

Sambuichi, H. 1966. Effects of radiation on frog eggs. II. Diploid and haploid eggs with heavily γ-irradiated cytoplasm. *Embryologica* 9:196–204.

Sanders, H. O. 1970. Pesticide toxicities to tadpoles of the western chorus frog *Pseudacris triseriata* and Fowler's toad *Bufo woodhousii fowleri. Copeia* 1970:246–251.

Saunders, J. W. and J. F. Fallon. 1966. Cell death in morphogenesis. In: M. Locke (ed.), *Major problems in developmental biology.* New York: Academic Press. Pp. 289–314.

Saunders, J. W., M. T. Gasseling, and L. C. Saunders. 1962. Cellular death in morphogenesis of the avian wing. *Dev. Biol.* 5:147–178.

Schaeffer, H. E., B. E. Schaeffer, and I. Brick. 1973. Effects of cytochalasin B on the adhesion and electrophoretic mobility of amphibian gastrula cells. *Dev. Biol.* 34:163–168.

Schmidt, G. A. 1933. Schnürungs- und Durchschneidungsversuche am anurenkeim. *Wilhelm Roux' Arch. Entwicklungsmech. Org.* 129:1–44.

Schultz, J. 1952. Interrelations between nucleus and cytoplasm: Problems at the biological level. *Exp. Cell Res., Suppl.* 2:17–43.

Schultz, J. 1959. The role of somatic mutation in neoplastic growth. In: *Genetics and cancer.* Austin: University of Texas Press. Pp. 25–42.

Schultze, O. 1895. Die Künstliche erzeugung von Doppelbildungen bei Froschlarven mit Hilfe abnormer Gravitationswirkung. *Wilhelm Roux' Arch. Entwicklungsmech. Org.* 1:269–306.

Seidel, F. 1932. Die Potenzen der Furchungskerne im Libellenei und ihre Rolle bei der Aktivierung des Bildungszentrums. *Wilhelm Roux' Arch. Entwicklungsmech. Org.* 126:213–276.

Shah, V. C. 1962. An improved technique of preparing primary culture of isolated cells from adult frog kidney. *Experientia* 18:239–240.

Shanmugaratnum, K. 1972. The pathology of nasopharyngeal carcinoma—a review. In: P. M. Biggs, G. de-Thé, and L. N. Payne (eds.), *Oncogenesis and herpesviruses.* Lyon: International Agency for Research on Cancer. Pp. 239–248.

Shea, M., A. L. Maberley, J. Walters, R. S. Freeman, and A. M. Fallis. 1973. Intraretinal larval trematode. *Trans. Am. Acad. Ophthalmol. Otolaryngol.* 77:OP784–OP791.

Sherbet, G. V. (ed.). 1974. *Neoplasia and cell differentiation.* Basal: S. Karger.

Shumway, W. 1940. Stages in the normal development of *Rana pipiens.* I. External form. *Anat. Rec.* 78:139–147.

Signoret, J. 1965. Transplantations nucléaires et différenciation embryonnaire. *Arch. Biol.* 76:591–606.

Signoret, J. and J. Fagnier. 1962. Activation expérimentale de l'oeuf de pleurodèle. *C. R. Hebd. Seances Acad. Sci.* 254:4079–4080.

Signoret, J. and J. Lefresne. 1970. Étude de l'apparition de l'asynchronisme dans la replication de l'acide désoxyribonucleique au cours du développement germe d'axolotl. *Ann. Embryol. Morphog.* 3:295–307.

Signoret, J. and B. Picheral. 1962. Transplantation de noyaux chez *Pleurodeles waltlii* Michah. *C. R. Hebd. Seances Acad. Sci.* 254:1150–1151.

Signoret, J., R. Briggs, and R. R. Humphrey. 1962. Nuclear transplantation in the axolotl. *Dev. Biol.* 4:134–164.

Simnett, J. D. 1964a. Histocompatibility in the platanna, *Xenopus laevis laevis* (Daudin), following nuclear transplantation. *Exp. Cell Res.* 33:232–239.

Simnett, J. D. 1964b. The development of embryos derived from the transplantation of neural ectoderm cell nuclei in *Xenopus laevis. Dev. Biol.* 10:467–486.

Simnett, J. D. 1966. Factors influencing the differentiation of amphibian embryos implanted into homologous immunologically competent hosts (*Xenopus laevis*). *Dev. Biol.* 13:112–143.

Simpson, N. S. and R. G. McKinnell. 1964. The burnsi gene as a nuclear marker for transplantation experiments in frogs. *J. Cell Biol.* 23:371–375.

Sládeček, F. and Z. Mazákova-štefanová. 1964. Nuclear transplantations in *Triturus vulgaris* L. *Folia Biol. (Praque)* 10:152–154.

Sládeček, F. and Z. Mazáková-Štefanová. 1965. Intraspecific and interspecific nuclear transplantations in *Triturus. Folia Biol. (Praque)* 11:74–77.

Sládeček, F. and A. Romanovský. 1967. Ploidy, pigment patterns and species specific antigenicity in interspecific nuclear transplantations in newts. *J. Embryol. Exp. Morphol.* 17:319–330.

Smith, B. G. 1912. The embryology of *Cryptobranchus alleghheniensis*, including comparisons with some other vertebrates. I. Introduction: The history of the egg before cleavage. *J. Morphol.* 23:61–158.

Smith, L. D. 1965. Transplantation of the nuclei of primordial germ cells into enucleated eggs of *Rana pipiens. Proc. Nat. Acad. Sci. USA* 54:101–107.

Smith, L. D. 1966. The role of a "germinal plasma" in the formation of primordial germ cells in *Rana pipiens. Dev. Biol.* 14:330–347.

Smith, S. W. 1950. Chloromycetin in the treatment of "red leg." *Science* 112:274–275.

Smith-Gill, S. J. 1973. Cytophysiological basis of disruptive pigmentary patterns in the leopard frog *Rana pipiens.* I. Chromatophore densities and cytophysiology. *J. Morphol.* 140:271–284.

Smith-Gill, S. J., C. M. Richards, and G. W. Nace. 1972. Genetic and metabolic bases of two "albino" phenotypes in the leopard frog, *Rana pipiens. J. Exp. Zool.* 180:157–167.

Soeiro, R., M. H. Vaughn, J. R. Warner and J. E. Darnell, Jr. 1968. The turnover of nuclear DNA-like RNA in HeLa cells. *J. Cell Biol.* 39:112–118.

Spalatin, J., R. Connell, A. N. Burton, and B. J. Gollop. 1964. Western equine encephalitis in Saskatchewan reptiles and amphibians, 1961–1963. *Can. J. Comp. Med.* 28:131–142.

Spemann, H. 1901. Entwicklungsphysiologische Studien am Triton-Ei. *Wilhelm Roux' Arch. Entwicklungsmech. Org.* 12:224–264.

Spemann, H. 1914. Über veizögerte Kernversorgung von Keimteilen. *Verh. Dtsch. Zool. Ges. Leipzig (Freiburg)* 24:216–221.

Spemann, H. 1919. Experimentelle Forschungen zum Determinations- und Individualitäts problem. *Naturwissenschaften* 32:581–591.

Spemann, H. 1921. Die Erzeugung tierischer Chimären durch heteroplastische embryonale Transplantation zwischen *Triton cristatus* and *taeniatus. Wilhelm Roux' Arch. Entwicklungsmech. Org.* 48:533–570.

Spemann, H. 1938. *Embryonic development and induction.* New Haven, Conn.: Yale University Press.

Špinar, Z. V. 1972. *Tertiary frogs from central Europe.* The Hague: Junk.

Spratt, N. T., Jr. 1971. *Developmental biology.* Belmont, Calif.: Wadsworth.

Srebro, Z. 1959. Investigations on the regenerative capacity of the diencephalon and the influence of its removal upon the development of *Xenopus laevis* tadpoles. *Folia Biol. (Krakow)* 7:191–202.

Stackpole, C. W. 1969. Herpes-type virus of the frog renal adenocarcinoma. I. Virus development in tumor transplants maintained at low temperature. *J. Virol.* 4:75–93.

Steen, T. P. 1968. Stability of chondrocyte differentiation and contribution of muscle to cartilage during limb regeneration in the axolotl (*Siredon mexicanum*). *J. Exp. Zool.* 167:49–78.

Stein, G. S., T. C. Spelsberg, and L. J. Kleinsmith. 1974. Nonhistone chromosomal proteins and gene regulation. *Science* 183:817–824.

Steinberg, M. S. 1957. A non-nutrient culture medium for embryonic tissues. *Carnegie Inst. Washingtpn, Yearb.* 56:347.

Stepina, V. N., V. E. Gourtsevicz, N. P. Mazurenko, N. M. Yarymova, M. M. Kaverznyeva, and Y. I. Lorye. 1976. Humoral antibodies to the capsid antigen of Epstein-Barr virus in Hodgkin's disease. *Neoplasma* 23:523–532.

Stevens, L. C. 1960. Embryonic potency of embryoid bodies derived from a transplantable testicular teratoma of the mouse. *Dev. Biol.* 2:285–297.

Stevens, L. C. and D. S. Varnum. 1974. The development of teratomas from parthenogenetically activated ovarian mouse eggs. *Dev. Biol.* 37:369–380.

Stroeva, O. G. and L. A. Nikitina. 1960. Nuclear transfer in amphibians and its significance for the differentiation problem. *Zh. Obshch. Biol.* 21:335–346.

Subtelny, S. 1958. The development of haploid and homozygous diploid frog embryos obtained from transplantations of haploid nuclei. *J. Exp. Zool.* 139:263–305.

Subtelny, S. 1965a. Single transfers of nuclei from differentiating endoderm cells into enucleated and nucleate *Rana pipiens* eggs. *J. Exp. Zool.* 159:47–58.

Subtelny, S. 1965b. On the nature of the restricted differentiation-promoting ability of transplanted *Rana pipiens* nuclei from differentiating endoderm cells. *J. Exp. Zool.* 159:59–92.

Subtelny, S. 1974. Nucleocytoplasmic interactions in development of amphibian hybrids. *Int. Rev. Cytol.* 39:35–88.

Subtelny, S. and C. Bradt. 1960. Transplantations of blastula nuclei into activated eggs from the body cavity and from the uterus of *Rana pipiens*. I. Evidence for fusion between the transferred nucleus and the female nucleus of the recipient eggs. *Dev. Biol.* 2:393–407.

Subtelny, S. and C. Bradt. 1961. Transplantations of blastula nuclei into activated eggs from the body cavity and from the uterus of *Rana pipiens*. II. Development of the recipient body cavity eggs. *Dev. Biol.* 3:96–114.

Subtelny, S. and C. Bradt. 1963. Cytological observations on the early developmental stages of activated *Rana pipiens* eggs receiving a transplanted blastula nucleus. *J. Morphol.* 112:45–59.

Subtelny, S., L. D. Smith, and R. E. Ecker. 1968. Maturation of ovarian frog eggs without ovulation. *J. Exp. Zool.* 168:39–48.

Summers, R. G. 1970. The effect of actinomycin D on demembranated *Lytechinus variegatus* embryos. *Exp. Cell Res.* 59:170–171.

Tarkowski, A. K. 1971. Development of single blastomeres. In: J. C. Daniel, Jr. (ed.), *Methods in mammalian embryology*. San Francisco: W. H. Freeman. Pp. 172–185.

Tartar, V. 1961. *The biology of Stentor*. New York: Pergamon.

Taylor, A. C. and J. J. Kollros. 1946. Stages in the normal development of *Rana pipiens* larvae. *Anat. Rec.* 94:7–23.

Thompson, D'A. W. 1942. *On growth and form*. New York: MacMillan Company.

Ting, H. 1951. Diploid, androgenetic and gynogenetic haploid development in anuran hybridization. *J. Exp. Zool.* 116:21–58.

Toivonen, S., D. Tarin, and L. Saxen. 1976. The transmission of morphogenetic signals from amphibian mesoderm to ectoderm in primary induction. *Differentiation* 5:49–55.

Toplin, I., P. Brandt, and P. Sottong. 1969. Density gradient centrifugation studies on the herpes-type virus of the Lucké tumor. In: M. Mizell (ed.), *Biology of amphibian tumors*. New York: Springer-Verlag. Pp. 348–357.

Townes, P. L. and J. Holtfreter. 1955. Directed movements and selective adhesion of embryonic amphibian cells. *J. Exp. Zool.* 128:53–120.

Tung, S. M. 1964. The developmental capacity of nuclei taken from the lateral mesoderm cells of *Bufo bufo gargarizans*. *Acta Biol. Exp. Sin.* 9:346–350.

Tung, T. C. and Y. F. Y. Tung. 1963. Nuclear transfers in vertebrates. *Acta Zool. Sin.* 15:151–167.

Tung, T. C., Y. F. Y. Tung, T. Y. Luh, M. Tu, and S. M. Tung. 1964. The developmental capacity of endoderm nuclei of *Bufo bufo gargarizans* as revealed by nuclear transplantation. *Acta Biol. Exp. Sin.* 9:217–223.

Tung, T. C., Y. F. Y. Tung, T. Y. Luh, S. M. Tung, and M. Tu. 1973. Transplantation of nuclei between two subfamilies of teleosts (goldfish—domesticated *Carassius auratus*, and Chinese bitterling—*Rhodeus sinansis*. *Acta Zool. Sin.* 19:201–212.

Tung, T. C., S. C. Wu, Y. F. Y. Tung, S. S. Yen, M. Tu, and T. Y. Lu. 1963. Transplantation of nuclei of fish cells. *Scientia* 7:60–61.

Tung, T. C., S. C. Wu, Y. F. Y. Tung, S. S. Yen, M. Tu, and T. Y. Lu. 1965. Nuclear transplantation in fishes. *Scient. Sin.* 14:1244–1245.

Tung, T. C., S. C. Wu, Y. C. Yeh, K. S. Li, and M. C. Hsu. 1977. Cell differentiation in ascidian studied by nuclear transplantation. *Scient. Sin.* 20:222–233.

Turpen, J. B., E. P. Volpe, and N. Cohen. 1975. The origin of thymic lymphocytes. *Am. Zool.* 15:51–61.

Tweedell, K. S. 1965. Cytopathology of a frog renal adenocarcinoma *in vitro* with fluorescence microscopy. *Ann. N.Y. Acad. Sci.* 126:170–187.

Tweedell, K. S. 1967. Induced oncogenesis in developing frog kidney cells. *Cancer Res.* 27:2042–2052.

Tweedell, K. S. 1969. Simulated transmission of renal tumors in oocytes and embryos of *Rana pipiens*. In: M. Mizell (ed.), *The biology of amphibian tumors*. New York: Springer-Verlag. Pp. 229–239.

Tweedell, K. S. and A. Granoff. 1968. Viruses and renal carcinoma of *Rana pipiens*. V. Effect of frog virus 3 on developing frog embryos and larvae. *J. Nat. Cancer Inst.* 40:407–410.

Tweedell, K. S. and D. C. Williams. 1976. Morphological changes in frog pronephric cell surfaces after transformation by herpes virus. *J. Cell Sci.* 22:385–395.

Tweedell, K. S. and W. Y. Wong. 1974. Frog kidney tumors induced by herpesvirus cultured in pronephric cells. *J. Nat. Cancer Inst.* 52:621–624.

Van der Hoeden, J. (ed.). 1964. *Zoonoses*. Amsterdam: Elsevier.

Verdonk, N. H. and J. N. Cather. 1973. The development of isolated blastomeres in *Bithynia tentaculata* (Prosobranchia, Gastropoda). *J. Exp. Zool.* 186:47–62.

Vogt, W. 1929. Gestaltungsanalyse am Amphibienkeim mit örtlicher Vitalfärbung. II. Gastrulation und Mesodermbildung bei Urodelen und Anuren. *Wilhelm Roux' Arch. Entwicklungsmech. Org.* 120:385–706.

Volpe, E. P. 1955. A taxo-genetic analysis of the status of *Rana kandiyohi* Weed. *Syst. Zool.* 4:75–82.

Volpe, E. P. 1961. Polymorphism in anuran populations. In: F. W. Blair (ed.), *Vertebrate speciation*. Austin: University of Texas Press. Pp. 221–234.

Volpe, E. P. and S. Dasgupta. 1962. Gynogenetic diploids of mutant leopard frogs. *J. Exp. Zool.* 151:287–301.

Volpe, E. P. and B. M. Gebhardt. 1965. Effect of dosage on the survival of embryonic homotransplants in the leopard frog, *Rana pipiens*. *J. Exp. Zool.* 160:11–28.

Volpe, E. P. and R. G. McKinnell. 1966. Successful tissue transplantation in frogs produced by nuclear transfer. *J. Hered.* 57:167–174.

Volpe, E. P. and J. B. Turpen. 1976. Massive colonization of the spleen by thymic cells. In: R. K. Wright and E. L. Cooper (eds.), *Phylogeny of thymus and bone marrow-bursa cells*. Amsterdam: Elsevier/North Holland. Pp. 113–122.

Wabl, M. R. and L. DuPasquier. 1976. Antibody patterns in genetically identical frogs. *Nature* 264:642–644.

Wabl, M. R., R. B. Brun, and L. DuPasquier. 1975. Lymphocytes of the toad *Xenopus laevis* have the gene set for promoting tadpole development. *Science* 190:1310–1312.

Waddington, C. H. and E. M. Pantelouris. 1953. Transplantation of nuclei in newts' eggs. *Nature* 172:1050–1051.

Wade, N. 1973. Microbiology: Hazardous profession faces new uncertainties. *Science* 182:566–567.

Wagner, E. K., B. Roizman, T. Savage, P. G. Spear, M. Mizell, F. E. Durr, and D. Sypowicz. 1970. Characterization of the DNA of herpesviruses associated with Lucké adenocarcinoma of the frog and Burkitt lymphoma of man. *Virology* 42:257–261.

Wallace, H. and T. R. Elsdale. 1963. Effects of actinomycin D on amphibian development. *Acta. Embryol. Morphol. Exp.* 6:275–282.

Wang, Y. T. 1963. Induced ovulation and maturation of the oocytes with human chorionic gonadotropin and steroid hormones in the toad, *Bufo bufo gargarizans*. *Acta. Biol. Exp. Sin.* 8:517–535.

Weed, A. C. 1922. New frogs from Minnesota. *Proc. Biol. Soc. Wash.* 35:107–110.

Weismann, A. 1892. *Das Keimplasma. Eine Theorie der Verebung.* Jena: G. Fischer. English translation by W. N. Parker and H. Rönnfeldt. 1915. New York: C. Scribner's Sons.

Weiss, P. 1939. *Principles of development.* New York: Henry Holt and Co.

Wier, K. A., K. Fukuyama, and W. L. Epstein. 1971. Nuclear changes during keratinization of normal human epidermis. *J. Ultrastruct. Res.* 37:138–145.

Wilde, C. 1960. Discussion comment. In: *Cell physiology of neoplasia.* Austin: University of Texas Press. P. 616.

Wildy, P., W. C. Russell and R. W. Horne. 1960. The morphology of the herpes virus. *Virology* 12:204–222.

Wilson, A. C., L. R. Maxson, and V. M. Sarich. 1974a. Two types of molecular evolution. Evidence from studies of interspecific hybridization. *Proc. Nat. Acad. Sci. USA* 71:2843–2847.

Wilson, A. C., V. M. Sarich, and L. R. Maxon. 1974b. The importance of gene rearrangement in evolution: Evidence from studies on rates of chromosomal, protein, and anatomical evolution. *Proc. Nat. Acad. Sci. USA* 71:3028–3030.

Wilson, E. B. 1893. Amphioxus and the mosaic theory of development. *J. Morphol.* 8:579–638.

Wilson, E. B. 1901. Experimental studies in cytology. I. A cytological study of artificial parthenogenesis in sea urchin eggs. *Wilhelm Roux' Arch. Entwicklungsmech. Org.* 12:529.

Wilson, E. B. 1903. Experiments on cleavage and localization in the nemertine egg. *Wilhelm Roux' Arch. Entwicklungsmech. Org.* 16:411–460.

Wilson, E. B. 1904. Experimental studies on germinal localization. I. The germ regions in the egg of Dentalium. II. Experiments on the cleavage-mosaic in Patella and Dentalium. *J. Exp. Zool.* 1:1–72, 197–268.

Wilson, E. B. 1928. *The cell in development and heredity.* 3rd ed. New York: Macmillan.

Wimber, D. E. and W. Prensky. 1963. Autoradiography with meiotic chromosomes of the male newt (*Triturus viridescens*) using H^3-thymidine. *Genetics* 48:1731–1738.

Witschi, E. 1929. Studies on sex differentiation and sex determination in amphibians. II. Sex reversal in female tadpoles of *Rana sylvatica* following an application of high temperature. *J. Exp. Zool.* 52:267–292.

Wolf, N. S. and J. J. Trentin. 1968. Hemopoietic colony studies. V. Effect of hemopoietic organ stroma on differentiation of pluripotent stem cells. *J. Exp. Med.* 127:205–214.

Wolf, O. M. 1929. Effect of daily transplants of anterior lobe of pituitary on reproduction of frog (*Rana pipiens* Schreber). *Proc. Soc. Exp. Biol. Med.* 26:692–693.

Wright, A. H. and A. A. Wright. 1949. *Handbook of frogs and toads of the United States and Canada.* Ithaca, N.Y.: Comstock.

Wright, P. A. and A. R. Flathers. 1961. Facilitation of pituitary-induced frog ovulation by progesterone in early fall. *Proc. Soc. Exp. Biol. Med.* 106:346–347.

Zambernard, J. 1973. In vitro growth of Lucké renal tumor cells. *J. Nat. Cancer Inst.* 50:577–578.

Zambernard, J. and R. G. McKinnell. 1969. "Virus-free" renal tumors obtained from pre-hibernating leopard frogs of known geographic origin. *Cancer Res.* 29:653–657.

Zambernard, J. and A. E. Vatter. 1966. The fine structural cytochemistry of virus particles found in renal tumors of leopard frogs. I. An enzymatic study of the viral nucleoid. *Virology* 28:318–324.

Zambernard, J., A. E. Vatter, and R. G. McKinnell. 1966. The fine structure of nuclear and cytoplasmic inclusions in primary renal tumors of mutant leopard frogs. *Cancer Res.* 26:1688–1700.

Ziegler, D. and Y. Masui. 1973. Control of chromosome behavior in amphibian oocytes. I. The activity of maturing oocytes inducing chromosome condensation in transplanted brain nuclei. *Develop. Biol.* 35:283–292.

Zoja, R. 1895. Sullo sviluppo dei blastomeri isolati dalle uova di alcune meduse. *Wilhelm Roux' Arch. Entwicklungsmech. Org.* 1:578–595.

zur Hausen, H. and H. Schulte-Holthausen. 1972. Detection of Epstein-Barr genomes in human tumor cells by nucleic acid hybridization. In: P. M. Biggs, G. de-Thé, and L. N. Payne (eds.), *Oncogenesis and herpesviruses.* Lyon: International Agency for Research on Cancer. Pp. 321–325.

INDEX

Index

313